AF699598

STORY OF THE
COMSTOCK

From gold dust and 'blue stuff' to the glamour of Paris and San Francisco, some prospectors had it all!

Silvia Anne Sheafer

Journal Publications P.O. 9750, Glendale, CA 91226

Second Printing - 2001

Printed in Glendale, California

Library of Congress Cataloging-in-Publication Data

ISBN 1-889971-00-6

A Milk Tin of Flaky Gold!

Swirls of hot wind flew off the dust, leaving a few flecks of yellow in the bottom of the tin milk pan. William Prouse, a young Mormon traveling on the wagon train going west, was amazed. He saw that they were gold - bright - yellow - glittering gold. It was unbelievable! Here was gold at the foot of Nevada's Sun Mountain. Yet the gold fields Prouse was heading for, were 150 miles west in California.

Prouse was nevertheless thrilled with his find. He ran back to the covered wagons and showed his companions. To them it had little significance. They saw only hardship and desolation. On the wild, burnt hills, Sun Mountain towered 5,000 feet above the desert, clothed with purple sage, rattle snakes, and jutting rocks. The canyon, and its twisting creek, where the wagons stopped, offered the pioneers nothing but a days respite, and water for the animals. It was El Dorado, over the Sierra Nevada, where the fabulous riches of the Mother Lode awaited to be unearthed.

Lacking imagination and foresight, Prouse had to agree with the others. He tossed away his specks of gold. So in the spring of 1850 the gold-seeking emigrants moved onward.

But lady luck was still beckoning. A spring snow blocked the passes over the Sierra summit. John Orr, the trail guide, gave orders to turn back to the valley of the Carson. Near the mouth of the canyon were Prouse had found his gold, at Spafford Hall's Station, the wagon train halted. The next day, mapping out his course, Orr marked the spot as Gold Canyon.

Panning for gold

While waiting for the Sierra snows to melt Orr went prospecting in the same gulch as Prouse. He trudged up the gorge to a point where the stone walls of the cliffs were barely wide enough to allow the sun to penetrate. Through the cut tumbled a creek of clear water. At its apex, Orr' spotted a slab of slate with a diagonal crack across its face. Pulling out his Bowie knife, he jabbed it into the crack, splitting the slate open. Water running over the slate, washed away the underlying dirt. And there it was again. A gold nugget the size of a hazelnut!

Excitedly Orr returned to camp and showed his piece of gold. The emigrants were no more interested, than they were over Prouse's discovery. After all if there was gold here, what must lay beyond the Sierra in the famous 49er gold fields? Orr pocketed his nugget. A few days later, when the Sierra was again passable, Orr and the others turned their back on Gold Canyon.

THE GREAT BASIN

Still more pioneer prospectors passed by way of Spafford Hall's Station.They too heard the miraculous tale of Orr's nugget.

That summer another wayfarer, working near where Orr had thrust in his knife, washed out 600 ounces of gold-dust in a single day. He sold it for $8,000, yet he journeyed onward. The gold was too flaky, too ragged. It's was alloyed with some fake "blue stuff" that choked the rockers and carried away the quicksilver.

There was pure gold in California! That was were they were going. By late August 60,000 emigrants passed by way of Stafford Hall's Station, non wishing to stay long enough to find the real treasure of the Comstock Lode.

Among the 60,000 who passed by Spafford Hall's Station that year was Horace Greeley, one-time editor of *The New York Tribune* newspaper, and author of "An Overland Journey from New York to San Francisco." He too was defiant, offering little hope for Gold Canyon. "Faster, faster!" he ordered the stage-driver. "This is desert, ought to be good for precious metal. It's not good for anything else." On to California he flew!

Along with the pioneers and news makers passed William Thorrington, James Fair, William Sharon, George Hearst. Some would return like Hearst, Fair and Sharon and become empire builders and state senators. Other miners, such as John Orr, settled on a farm in Sonoma County, California, the gold nugget his single claim to prospecting. Prouse eventually returned to the land around Salt Lake City, Utah.

Although the two gold discoveries were not sufficient to stay the progress of the Orr wagon train, the news was significant to pass back along the trail. It was printed in the *California Courier* of July 8, 1850 and attracted California miners back to Gold Canyon in the fall of 1850.

Return of the Gold Seekers

The following spring trekked more emigrants, miners, traders and their pack trains. A party of Mormon farmers settled in the fertile valley of the Carson and named it Mormon Station. But there was no water for their crops. Drought was the curse of Washoe - a region named for the Indian tribe that roamed there. The Mormons imported a hundred Chinese to dig an irrigating ditch from the Carson River to their farms. When the placers in Gold Canyon grew richer and richer, Chinese and missionaries abandoned the ditches for the yellow ore.

Chinese miner at his rocker.

Vigilante Justice

Tempers flared between the two groups. When one Chinese was accused of infringing on a Mormon's claim, he was strung up by his long, back pigtail as a warning to others to be wary. To protect themselves the Chinese built their cabins around the mouth of the creek, and Spafford Hall's Station became Chinatown.

After a profitable winter in the California placers, a motley bunch of individuals joined the Mormons. William "Lucky Bill" Thorrington, returned and made good in the canyon placers. He gained control of the Carson Canyon Toll Road, and built a way station, collected toll fees, and dispensed charity to those in need.

James Fennimore, a Virginian, had killed a man in California. To escape punishment, he fled to Utah Territory, then including part of Nevada. He changed his name to Finney and was tagged with "Virginny," and "Old" due to his capacity for liquor. He carried a gun, a

A motley group of miners

bottle in his pocket, and balanced a shovel on his shoulder. Finney became a permanent fixture at the Spafford Hall's bar.

Into the camp came an overbearing Canadian herder, with his flock of sheep. Henry Thomas Paige Comstock, son of Old Noah Comstock of Cleveland, Ohio, claimed to be "a regular born mountaineer." Others called him a "natural born liar."

One of the first women to enter Gold Canyon was Eilley Orrum, who was said to have divorced two husbands, both Mormons, and one Mormon bishop. Satisfied with being single once more, she became mother to all the miners. In doing so she settled at a place called Johntown, a camp that had sprung up four miles up the canyon from Spafford Hall Station.

Eilley built a log cabin, took in boarders, and let it be known that she would not "sleep" them. Eilley was a good cook. Her pork and beans and batter biscuits soon became as prominent as her boarders. Known as"Johntowners" they included, "Old Pancake" Comstock, "Old Virginny," "Sandy" Bowers, Pat McLaughlin and several others. When the creeks began to run low, Eilley packed up her boarders laundry, and headed for the hot-water springs in Washoe Valley.

Although the Johntowners ate with Eilley, their entertainment was in Chinatown.

This prospector takes time out to count his pieces of ore.

Over green baize tables, they "bucked the tiger and wrestled with the beasts of the jungle," squandering the gold-dust they had found during the day. Saturday nights, upstairs at Hall's Station, the miners forsook Monte for dancing. The "yaller-backed fiddle" struck up at eight o'clock, and the call was "Take your partners for the French-four." Women were scarce at Spafford Hall's. There was always a scramble for Eilley Orrum, Laura Ellis, and "Princess Sarah," daughter of Winnemucca, chief of the Paiutes.

Fireworks often ended the Saturday night festivities. After heavy drinking fights broke out between the miners. One time the Paiutes stampeded their mustangs. An Indian and a miner were occasionally killed in the melee. But manslaughter was nothing in comparison to thieving. Claim jumping and cattle rustling were the most heinous of crimes. You could steal a man's wife, but you couldn't steal his ox nor anything else.

According to Myron Angel's *History of Nevada*, the Johntowners sliced off the ears of horse-thieves, and sometimes his nose. Once they hanged an innocent man named Snow for a thieving crime, committed by a man known as Edwards. Snow had vehemently denied the crime. Nevertheless they dropped a noose over his head, and hoisted him off the ground. When he had strangled a bit, the rope was eased and the Snow was given a chance to further confess. But Snow commenced to curse and was strung up again. On the third suspension the vigilantes "kept him hanging too long." Snow was dead.

Lucky Bill

Edwards fled to the sanctity of Bill Thorrington's. "Lucky Bill" was hauled into court and accused of shielding a horse-thief. He was quickly found guilty and sentenced to hang. He was strung up to a cottonwood tree were he was heard singing, "The Last Rose of Summer."

"Blue Stuff" in the way of gold

All summer the miners worked knee-deep in the cold mountain water, rocking their cradles, and bemoaning the "blue stuff" that clogged their riffles. By handfuls they angrily tossed the cobalt mass as far down the gulch as they could throw. They further observed that the higher into the canyon they worked the bluer and heavier became the "stuff," while the gold grew pale and lighter. In the beginning the gold brought $14 to $15 an ounce. Now, with the "blue stuff" the price dropped to $11 an ounce.

No one thought to examine the "blue stuff." It wasn't until the spring of 1851 when two brothers, Allen and Hosea Grosch, arrived from New York that the "blue stuff" was seriously observed. In their saddle bags the brothers brought several technical books on mining and on chemistry. Methodically they began to mine for gold. Before long they too found their riffles clogged with the "blue stuff."

Instead of tossing it away, they took it back to their cabin, wet it, and ground it to a fine powder. Then they baked it in an improvised oven. In the button of the pot they found a dark colored button. Could it be silver! They had to test it further. They dropped the button into a beaker with nitric acid, then held the beaker over a flame. Slowly the button disappeared in the acid solution. The next day they wrote their father about their discover. "It resembles thin sheet lead - broken very fine." Allen and Hosea now faced a dilemma. Should they tell the other miners the nature of the material they were tossing away? They decided not too. They would first develop their own claim.

The Grosches worked their claim all summer. In the winter they went to Mud Springs, El Dorado County, California. There, they searched for gold, trying to earn enough money for the Nevada claim. Returning in the fall of 1855 they dug up the canyon and discovered a spring with a great vein that sparkled with gold and held the "blue stuff" - bluer than ever.

Eilley Orrum

While in El Dorado the Grosches had organized a stock company. On their return they were accompanied by Richard Bucke, a young Canadian who worked for them. Bucke had been told nothing about the "blue stuff." The Grosches confided in their only partner George Brown - station master at Gravelly Ford at the sink of the Humboldt. He promised them all his savings, about $600, to develop the lode. They told one other person, Eilley Orrum, about the claim - but not the "blue stuff"," and swore her to secrecy. For her kindness, the Grosches promised her a claim right next to their own - the "Pioneer."

The brothers built a stone cabin, filled it with chemicals, assaying apparatus, scientific books, a cupel muffle-furnace with canvas bellows, and a work bench. They worked their claim, gold-washing in the canyons and on the hillsides. They wrote that the first assay yielded returns of $3500 per ton! It was fabulous!

Still they needed more money and equipment. They wrote to Brown, urging him to join them. He promised to come.

Could this be the Grosches great silver discovery?

Cradle and Riffle Bars

Panning for gold became too slow for the miners.They needed a faster way to wash the gold from the gravel. "Cradles" were made of hollowed longs or a cradle cut in half, open at the end. On top of the cradle the miner nailed a box. This was called a "hopper." On the bottom of the hopper was a piece of iron with small holes. The holes screened the gravel. Little wooden "riffle bars" were laid on the lower end of the cradle, catching the gold as it washed through. Along the side of the cradle were rockers like those on a child's cradle. A handle was put on the cradle to it could be rocked. Two men could work a cradle.

Gravel was then shoveled it into the cradle and water poured over it. As the cradle rocked pieces of gravel broke into smaller pieces. The gold fell to the bottom caught between the riffle bars. Here, the "blue stuff" was clogging the extraction.

Curse of the "Blue Stuff"

On a warm August afternoon Hosea accidentally cut his foot with his pick. A minor injury and one that Hosea ignored. A red line soon raced up his leg, and fever raged through his body. A day later he began to feel better. Hosea sat outside the cabin and was greeted by Laura Ellis. She brought bad news - George Brown had been murdered by thieves. The brothers were crest-fallen.

Then Laura asked them if they were sure of the "blue stuff," she would devote $1500 to their claim. Allen assured her it was. He gave her a piece of blue quartz, thanked her, and put her name down for three hundred feet of the claim. Afterwards he recorded the notation in his ledger, and pointed up the hill towards the prospected area. A few days later tragedy struck once more. Hosea had died of locked jaw.

Overcome with grief, Allen swore to go on - for Hosea's sake. He wrote his father and told him of Hosea's death, and that he now fell heir to his dead son's share of the claim.

Discovery of the Comstock Lode June, 1859. Comstock is standing center.

Allen Grosch was anxious to return to California in the fall for he still needed money to develop the claim. He packed up his charts and notes, placing them in a box then nailing it shut. Before leaving for the winter Allen had to find someone who would care for its precious contents, and watch over his cabin. Allen choose "Old Pancake" Comstock, and promised if he would take good care of the cabin and its contents, upon returning in the spring he would give him one-fourth interest in his claim. Agreeing, Allen had Comstock signed to the provisions.

Misfortune seemed to follow Allen Grosch. That winter he and Bucke were trapped in a Sierra blizzard. When they were at last found by a party of miners their legs were badly frozen. Gangrene was feared. Quickly a saw was found and one of Bucke's legs and his other foot were amputated. Allen heard Bucke's struggle and refused amputation. He died soon after.

Fools of Fortune

Comstock learned of Allen's death the following spring. All the curiosity he'd had about the locked box, came forefront. Grabbing his axe he knocked off the lid. Books, equipment, and blue rocks with white lines darting through them sparkled in the light. Unable to recognize the significance of its contents, he wildly tore all apart. "Iron pirates!" he yelled, breaking everything he could lay hands on, tossing maps, notes and papers into the fire. In a fit of rage he stuffed the red and blue rocks in his pocket, and set out to find **HIS** claim.

For weeks Comstock scoured the mountains, searching for the Grosches' claim. His actions caught the attention of another miner. "Old Virginny" decided to investigate Comstock's activities. A fresh outcropping caught his eye. On a piece of yellow paper, "Old Virginny" signed James Fennimore, February 22, 1858. Although Fennimore never developed the claim it became known as "Old Virginny's" ledge.

Comstock - Like a sentry

Fearing the Fennimore had stolen his property, Comstock staked out a claim - for ranching purposes - on every likely spot on the mountain. Yet, Comstock still didn't know what the red rock nor the "blue stuff" was.

Comstock finally sent a sample of the rock to a Grass Valley metallurgist. Before it could be identified, the metallurgist was murdered. Next, a sample was sent to a geologist in Placerville who said it looked like blue limestone. He suggested it be sent to Sacramento. There, the response was that the rock was worth $2200 to the ton in gold. Again, however, the "blue stuff" was not identified.

Each day Comstock grew with near paranoia. He carried a shotgun, and like a sentry paced his claims. Then it happened. On a warm day in January, the walls of Gold Canyon echoed with wild shouts of joy. Part of the lost Grosch claim was rediscovered. "Old Virginny" had plunged his hand in a gopher hole, and pulled out gold. Unknowingly he had unearthed the Gold Hill outcroppings of the famous Lode. Prospecting with him was "Big French John," Aleck Henderson, and Jack Young. All claimed ownership on what Comstock said was rightfully his. All staked out their claims. Comstock was forced to take what was left.

News of the gold strike spread like wild fire. Eilley, with the aid of her boarders, Jim Rogers, Joe Plato, and Sandy Bowers, packed her boarding house on a pack train and headed up Gold Canyon. There she had it set up, with its walls abutting on "Old Nick's" Bar. When Rogers failed to pay his board bill, Eilley took over his claim.

Gold filled the miner's rockers, and each day they cursed the "blue stuff" they tossed away. And still, the claim the Grosches had secretly kept, baffled Comstock. Where was their treasurer?

About this time Comstock learned that two other miners, Pete O'Riley and Pat McLaughlin, were mining over in Spanish Ravine at the head of Six-Mile Canyon. This was where Comstock had staked off one of his claims for grazing purposes. He decided to take a look. In June of 1859 the second gold discovery occurred.

Measuring gold by the troy ounce - 1860s

Comstock couldn't believe his eyes. Here was a mound off gold-dust - the first output of the famous Ophir mine at the opposite end of the Lode. He fell to his knees, scooping up gold, and the blue earth that sparkled like diamonds. Here was the lost ledge of Hosea and Allen Grosch. Excitedly he struggled to his feet. "You have struck it, boys!" Then he coldly informed the two astounded miners that they were trespassing on his land. Comstock said that he and Manny Penrod had located the claim last winter. He neglected to tell them he had also sold a tenth to "Old Virginny."

Skulduggery

Comstock, in rage accused O'Riley and McLaughlin of jumping his claim - a heinous crime. He demanded the claim. To the dismay of all, rather than continue the row, the two men yielded. Finally Comstock demanded one hundred feet for water rights, knowing the claim was useless without water. All that now stood in Comstock's way was the of the one tenth belonging to "Old Virginny."

That night, before news of the gold strike reached Johntown and "Old Virginny," Comstock formulated a devious plan. That night he met "Old Virginny" in Spafford Hall's bar With a bottle of "tarantula juice," and a bobtailed mustang, the exchange took place. When "Old Virginny" sobered up, he often lamented, that he was the most startling example of magnificence and misery the world had ever seen. Why, because he was able to ride a $60,000 horse and yet had to ride him bareback because he could not scare up cash enough to buy a saddle.

Within a week Comstock formed the firm of Penrod, Comstock & Company. Emanuel Penrod had claimed the land earlier and was a friend of Comstock. The mine was later known as the Ophir - the first claim recorded on the Lode. Within one year the Ophir stock went up to $3000 a foot. Thereafter Comstock referred to his claim as the "Comstock Lode."

The discovery did not create an immediate rush to the district. It did attract the attention of J.F. Stone, a partner in the Stone and Gates trading post on the Truckee River. Stone obtained samples of the ore, passed them on to J.J. Ott, who gave samples to Judge James Walsh, a prominent mill and mine owner and Melville Atwood, both of Grass Valley, California. The Atwood assay showed a result of $3,876 in gold and silver per ton at a time when any gold and silver ore which assayed over

Over the Sierra Rush to Washoe!

$100 per ton was considered a good mining prospect. The assay results sent Judge Walsh and his partner, Joseph Woodworth, hurrying to Washoe. After the *Nevada Journal* published the story on July 1, 1859, close on their heels came dozens of other prospectors from the Nevada City - Grass Valley area.

At the same time riches were not proving enough for "Old Pancake" Comstock. He staked out other claims, and brazenly alluded that they were his, whether or not they were. He gave pans of gold-dust to visiting ladies, and induced one, the wife of a poor Mormon, to marry him. To satisfy the husband, Comstock gave him a horse and threw in sixty dollars. When his bride ran off with another man, Comstock brought her back and locked her in an upstairs room. Sometime later when he returned to the room, he found the window open. His wife had run off with a passer-by.

California miners return to Nevada

By now the same madness which had caused others to leave home for their "pot of gold" now seized the new Californians, sending them back over the Sierra in what became known as the "rush to Washoe." Make-shift communities, such as Gold Hill, and Silver City soon were born.

Birth of a City

On a moonlight night in October of 1859, Comstock, and "Old Virginny," with a bottle snugly tucked under his arm, were leading some other miners into the purple sage. All were mellow with whiskey. Suddenly "Old Virginny" tripped, and his bottle went flying, hitting a rock and spilling its contents. Comstock laughed with glee. Not so "Old Virginny" who picked up the bottle and swung it to and for, solemnly proclaiming, "I baptize this spot Virginia Town."

As the summer wore on, the miners soon spoke of their wretched, dusty camp and its canvas shacks as a city. Thus Virginian Town became Virginia City.

Forming Gold Districts

Mining districts were a product of the California gold rush. They were enacted by the miners to protect and legalize their mining claims and transfers. There were no such laws in effect in the Utah Territory area surrounding Virginia City.

Confusion and lawlessness had intensified with the huge discoveries of silver and gold. Miners often found themselves outside of effective government. They had no legal means of holding and taking their claims. To protect themselves they adopted the mining district idea of California. This worked in some locations, but without a state government to back them up, there were still problems with filing claims, and claim jumping.

News of the gold strike in Virginia City traveled far and wide. All spring and summer long prospectors began streaming into the area. At one time there were over 350 wagons, coming east and crossing the Sierra Nevada range, miners leaving the Mother Lode of California to stake a claim in the rich fields of the western Utah Territory.

While miners fought to protect their claims, the Chinese prospectors panned the tailings.

Birth of The Territorial Enterprise

Making new laws and regulations was important to a growing community. So too was a newspaper. While some pioneers sought gold and silver, others desired a creative line of work. As luck had it, two brothers W.L. Jernegan and Alfred James were about to give birth to a long, and lasting enterprise.

If one were to hear them on that cold December day in 1858, they might not have believed such an endeavor as a newspaper was about to come off the press. At Mormon Station two bearded men, assisted by an apprentice boy, wrestled with a secondhand Washington printing press. The cannon-ball stove in the corner of the drafty shack may have glowed red with fire, but it hardly made a dent in the antarctic cold that enveloped the entire territory of western Utah.

No one knows why the two men would bring their press overland from Salt Lake City to Mormon Station, the community which later changed its name to Genoa with a community of fewer than two hundred residents. What ever the reasons, here they were about to produce the regions first weekly newspaper. As it happened the paper was named by a miner friend in California. Washington Wright suggested that since the venture was a new enterprise, in the then Territory of Western Utah, what could be more appropriate than The Territorial Enterprise.

Like the prototype newspaper, printers Jernegan and James wore ink-stained frock coats, dented top hats, with breath perfumed by strong waters. They carried a type stick in one hand, the other on the stock of a belted gun. Facing Indians, the wilderness, drunken and disgruntled readers and creditors was just part of the newsmans daily job.

After the first edition rolled off the presses it was christened, quite naturally, by everyone in the nearby Stockade Bar. That winter snows up to one-hundred inches fell on

Territorial Enterprise.

VOL. 1. CARSON CITY, NEVADA TERRITORY, SATURDAY, DECEMBER 17, 1859. **NO. 45.**

Territorial Enterprise.

Published every Saturday morning, at the Office on Carson st., between 2d and 3d, up stairs, Carson City, Nevada Territory.

—BY—

W. L. JERNEGAN & CO.

Terms,—Invariably in Advance,

One Year	$5 00.
Six Months	4 00.
Three Months	3 00.
Single Copies	25.

Advertising.

One Square, 10 lines, first insertion, $3—each subsequent insertion, $1 50. Business Cards of ten lines or less, one year, $25; three months, $10.

A liberal discount will be made on the above rates, for all yearly and quarterly advertisements which exceed one square.

Agents for the Enterprise:

THOMAS BOYCE	SAN FRANCISCO.
CHAS. A. CRANE	SAN FRANCISCO.
A. BADLAM & E. B. DAVIDSON	SACRAMENTO.
E. M. HILL	MARYSVILLE.
J. C. KERLEY	PLACERVILLE.
J. M. CLARK	COLOMA.
W. S. DAY	DIAMOND SPRINGS.
JOHN W. GREAR	DOWNIEVILLE.

Departure of Col. Musser.

Col. J. J. Musser, our recently elected Delegate to Congress, left for Washington, on business connected with the interests of Nevada Territory, on Monday morning last. Col. Musser bears with him the confidence, respect and hearty good will of our people. His abilities are undoubted, and his opportunities for earning a just fame are as fair as any now presented. He has reasons numerous and cogent to urge upon Congress, in favor of Nevada Territory. He possesses the requisite ability to present them properly; and if Congress fails to give us what we so much need, a separate organization, it will not be for want of energetic representation of our interests. We say, God speed! to Col. Musser. It will be a proud day for him, and for the residents of these Valleys, when he shall return with the Bill organizing the Territory of Nevada. And it will be a no less proud day for our

Ourselves.

One year ago to-day, the first number of the TERRITORIAL ENTERPRISE was issued at Genoa. Our publishing room was in Singleton's Hall, Nevada Hotel, a room indiscriminately used by preachers, debating-clubs, secret societies, and once at least, for a prison. Upon the latter occasion we had a man accused of crime chained to our printing press, with a log chain, for two days and a half. What secrets that old Hall might tell, could it, by chance, be endowed with the gift of speech! Our establishment was removed to Carson City, on Thursday, Nov. 10, 1859. We now occupy half of the upper part of Major Ormsby's adobe building, southwest corner of the Plaza. Our volume is not yet completed by seven numbers, owing to the fact that we have twice been compelled to suspend our issue for want of paper. During last winter, most of our paper was brought over on snow shoes, by attaches of this office. Many a time in the past year have we suffered for lack of fuel, and been pinched for want of actual necessaries of life. But so far we have struggled on successfully, and to-day we find ourselves in more comfortable circumstances in many respects. To be sure we still have to descend from the editorial tripod to superintend the cooking of a beefsteak, the seasoning of a bean soup, or the concoction of a pot of coffee, (and in this line, let us indulge our vanity by saying that we yield to none other as a caterer to the *physical* man, whatever our shortcoming in regard to intellectual pabulum.) But as we have said, we feel that we have made a step in advance. It has been and still *is* our aim, to lay the foundation of a reliable newspaper, such as the rapidly increasing population, and the developing interests of our pet New Territory demand. So far as literary ability is concerned, we make no claims for the past. The public can readily judge how far we *could* devote ourselves to the nice details of "fine writing," un-

PR

To the
incl
ries
of Ne

Havin
Executi
Governm
deeming,
the subj
curse of
to you m
of the P
at the p
At th
semble,
deavor
rule of A
for life,
vain pet
the unjus
mons to
obnoxio
no Cour
those co
of the S
cal right
certain
were op
our Co
ballot-b
Under
ored to
positio
sion
assu
prop
all nece
mation
Congres
tection.
Since
prived, I
of our
James M
were dev
people,
grave in
who sho
stone,
of God
With
tempt f
baugh,
Court in
us as be
ble stam
we at

Mormon Station. Drifts of incalculable depth blocked the high passes of the Sierra. Twice the paper stock ran out and the brothers were unable to find newsprint to meet their publication date. Thus at the end of their first year their volume was incomplete by six papers.

Money to operate the newspaper was often hard to come by. Sometimes subscribers paid in trade goods. If half a bear, a hind quarter of venison, or a dozen sage hens arrived in payment of the delinquent subscription, the staff lived high on the hog. Yet despite the inconvenience of money and inclement weather The Territorial Enterprise survived.

In its earliest years the front-page date line read "Carson Valley, Utah Territory - Published Every Saturday Morning at the office on Mill Street, Genoa, Carson Valley."

First office of The Territorial Enterprise - 1862
Courtesy of the Nevada Historical Society

The newspaper was destined to occupy many premises in the next hundred years. The brothers moved from the one-story wooden shack to quarters in Singleton's Hall, where many wild adventures took place. At one time Mormon Station had no jail. On an occasion, a prisoner awaiting trail was chained to the Washington hand press with a logging chain for three days. Local news was colorful as well: births and deaths, the departures of the great Overland Stages for Sacramento and Salt Lake City, disquieting stories of settlers and Paiute Indian events. shootings and hangings, discoveries of gold and silver, sales of mine stock, advertisements, theater announcements, and community doings.

National news arrived by way of six-week old copies of the Boston Advertiser, New York Tribune, New Orleans Picayaune, and other leading journals of news and opinion, lifted as was then the custom.

A description of the pioneer newspaper's first years was presented in Harper's Monthly by J. Ross Browne, staff reporter and sketch artist.

"Chief among the curiosities of Carson City is The Territorial Enterprise, a newspaper of an origin anterior to the mining excitement. I was introduced to "the Colonel" who presides over the editorial department and found him uncommonly strong in the ultimate destiny of Carson. His office was located in a dirty frame shanty, and general rubbish of dark and literary aspect; there those astounding editorials which now and then arouse the public mine are concocted. The Colonel and his compositors live in a sort of family fashion, entirely free from the rigorous etiquette of such establishments in New York. They cook their food in the composing room (which is also the editorial and press room) and being, as a general thing, short of plates, use the frying pan in common for that purpose. In cases of great festivity and rejoicing, when a subscriber has settled up arrears or the cash is paid down for a good job of hand-bills, the Colonel purchases the best tenderloin steak to be had in the market, cooks it with one hand while with the other he writes a letter of thanks to the subscriber or a puff on the hand-bill."

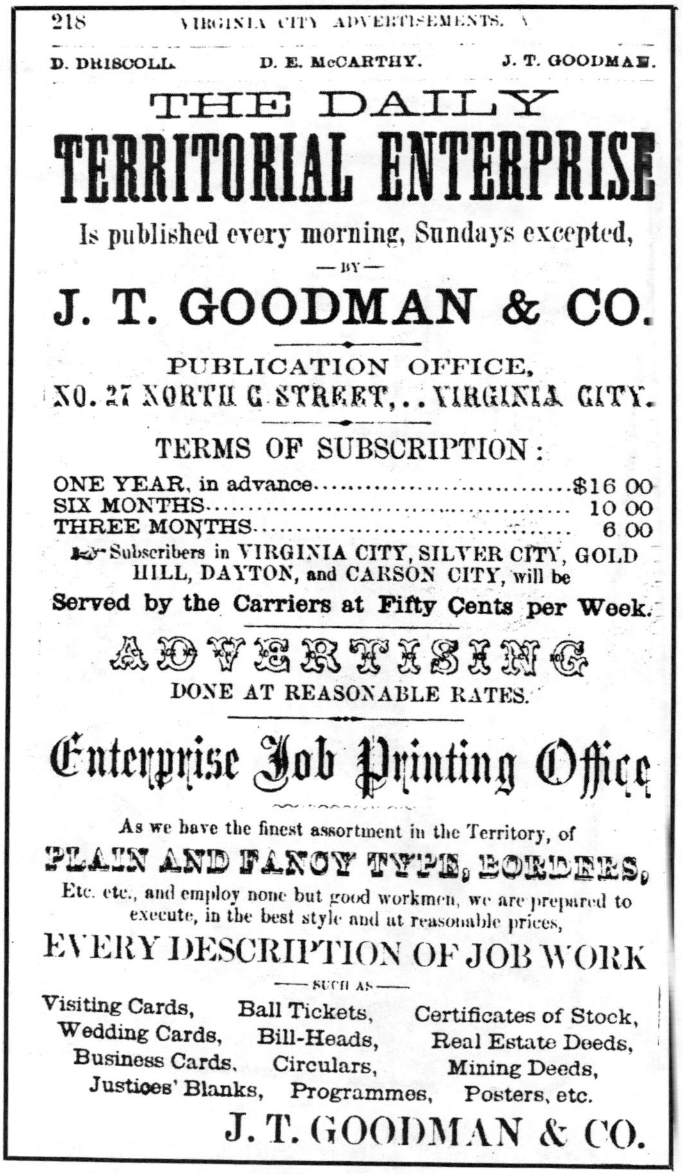
218 VIRGINIA CITY ADVERTISEMENTS.

D. DRISCOLL. D. E. McCARTHY. J. T. GOODMAN.

THE DAILY
TERRITORIAL ENTERPRISE
Is published every morning, Sundays excepted,
— BY —
J. T. GOODMAN & CO.

PUBLICATION OFFICE,
NO. 27 NORTH C STREET,.. VIRGINIA CITY.

TERMS OF SUBSCRIPTION:

ONE YEAR, in advance........$16 00
SIX MONTHS........10 00
THREE MONTHS........6 00

Subscribers in VIRGINIA CITY, SILVER CITY, GOLD HILL, DAYTON, and CARSON CITY, will be Served by the Carriers at Fifty Cents per Week.

ADVERTISING
DONE AT REASONABLE RATES.

Enterprise Job Printing Office

As we have the finest assortment in the Territory, of
PLAIN AND FANCY TYPE, BORDERS,
Etc. etc., and employ none but good workmen, we are prepared to execute, in the best style and at reasonable prices,
EVERY DESCRIPTION OF JOB WORK
— SUCH AS —

Visiting Cards, Ball Tickets, Certificates of Stock,
Wedding Cards, Bill-Heads, Real Estate Deeds,
Business Cards, Circulars, Mining Deeds,
Justices' Blanks, Programmes, Posters, etc.

J. T. GOODMAN & CO.

Newspaper advertisement

During the ensuing years The Enterprise had a succession of new owners. In August of 1859 the James partnership was acquired by Jonathan Williams and the firm name changed to Jernegan & Co. A year later the newspaper appeared without Jernegan's name and listed Williams & Degroot as editors and proprietors. On September 29, 1860, Williams' name alone adorned the masthead, and the office was moved to Virginia City.

While ownership changes were taking place, another significant event was occurring. **Samuel Langhorne Clemens** arrived in the territory, and assumed the pen name of Mark Twain. It was here in the raw and Wild West he originated his humorous style of treating wild exaggerations with absolute seriousness.

Clemens was born in Florida, Missouri in 1835. A few years after he was born the family moved to Hannibal, Mo., a port on the Mississippi River. When he was twelve his father died. Samuel left school to become a printer's apprentice. For a time he worked on his brother's newspaper, writing humorous stories and verse. Then he became a "tramp printer," working in New York City, Cincinnati, and St. Louis.

Samuel Clemens acquired his pen name "Mark Twain" while a river pilot (the pilot's call that the river was two fathoms deep). With the beginning of the Civil War, Twain joined the Confederate Army. Two weeks later, however, he managed to get discharged. Twain then set out for Nevada. In 1862 he began his literary career as a reporter on The Territorial Enterprise.

Mark Twain

It was at the fledgling newspaper Twain originated his humorous style of treating wild exaggerations with absolute seriousness. He served on The Enterprise staff with Joe Goodman, and Dan De Quille. The newspaper was well-known for its hoaxes, and Twain was not a reporter to shy away from such shenanigans .

Twain's most celebrated hoax was "The Empire City Massacre." Printed in The

Enterprise with a wealth of substantiating detail, the story told how wholesale murder had been committed by one "P. Hopkins or Philip Hopkins'" at Empire, the stamp-mill city near Carson.

There Hopkins allegedly killed his wife and seven children, maimed two others, and severed his wife's head . The fugitive then fled to Carson City where he cut his throat, and fall dead on the the sidewalk in front of the Magnolia Saloon. Twain quoted his informant as Old Abe Curry, Carson's first citizen and patriarch. Curry blamed Hopkins' fatal derangement on unfortunate speculations in Comstock mining properties resulting in reports that had appeared in San Francisco newspapers. "It is presumed that this misfortune drove him mad and resulted in his killing his family," Curry was quoted as saying.

For local residents the story was indeed a hoax. Hopkins, owner of the Magnolia Saloon, was unmarried. Other flagrant inconsistencies were recognized as merely monumental fabrications in bad taste.

However so sensational and realistic was the account that the Gold Hill News in California promptly lifted the story for its own edition. The gruesome account was readily believed by its readers. The next day The Enterprise printed a retraction, while competing newspapers joined the News in denouncing the affair for its shockingly bad taste.

A year later De Quille assigned Twain to cover the Nevada Legislature at Carson City. Twain, now a statehouse reporter, worked for the most influential paper in the territory. His writings began to take on loftier aspects than the hassles and hoaxs of Virginia City.

A few years later Twain moved to San Francisco. Working as a reporter there, his story, "The Celebrated Jumping Frog of Calavares County," made him widely known.

Twain's masterpiece was *Huckleberry Finn*, often considered as the first great American novel, and the forerunner of the realistic novel. He shaped the American language with great artistry, and used the Far West and the Mississippi as almost his only source of material.

Twain wrote a wide variety of books including, *The Prince and the Pauper*, *Connecticut Yankee at King Arthur's Court*, *The Mysterious Stranger*, and *The Innocents Abroad.*

Twain was basically not a very well educated person. But his great wit and charm made him popular in the best drawing rooms in America and Europe. He died in 1910.

Opening Night

"Miss Menken, Miss Morris, Little Miss Crabtree, five minutes to curtain!" The excitement of the callboy, making his rounds of dressing rooms, fore told the nightly theatrical performances.

Out front crystal chandeliers sparkled. The orchestra could be heard tuning up. Opening night stage boxes were reserved for the likes of John MacKay and Adolph Sutro. Virginia City's rich and powerful were present with their ladies on their arms and diamonds as big as robins' eggs pinned to their starched shirt bosoms. The gamblers were there too, without ladies, but with diamonds studs in their shirt sleeves somewhat bigger than those of the Comstocks aristocracy.

Soon the velvet curtains would rise. The pretty and talented young women would grace the stage of one of several of new, elegant Victorian theatres. Thus was the elegance and extravagance of Virginia City theatres.

The prospectors loved their musical halls and opera house. So much so that only a few years after the discovery of gold and silver, there were entertainment houses a plenty. And there were as many beautiful performers to enhance the fabulous interiors.

In 1863 Maguire's New Opera House was the first to open. Owner Tom Maguire, who had other theaters throughout the gold areas of California and the Utah Territory, had commissioned a magnificent stage curtain depicting a sunset at Lake Tahoe. It was reputed to have cost millions. Actually it was painted by Jacob Wimmer, an Italian artist as settlement for an overdue bill.

A double tier of boxes flanked the spacious stage with chairs upholstered in red velvet. The rails were swathed in matching velvet. Red velvet cords draped across the aisles, separating the box seats from general admission. Glittering French chandeliers hung from the ceiling, and a lush Turkey carpet clothed the floor. Handsome spittons replaced the accustomed sawdust.

The premiere production was "Money." It starred Julia Dean Hayne and Walter Leman. However, before the curtain went up, a shooting interrupted the evening entertainment.Two locals gamblers on bad terms, found themselves seated next to each other. Unable to resist the temptation, revolvers were drawn and shots fired. With the flash and bang of gun powder, crystals shattered and ladies screamed. As the contestants paused to reload, Maguire's ushers threw them out. Moments later the curtain was rung up.

In the ensuing years there would be more gun fights and new theatres when John Piper entered the scene. Without performers, however, there would have been no elegance or stylish attired gentry. The women who made the footlights shimmer were not only beautiful, but adventurous. Some of the actresses even became millionaires. A glance at these enduring ladies is certainly an important chapter in the history of the Lode.

Ladies of the Theatre

Adelina Patti

Adelina Patti

Adelina Patti (1843 - 1919), was said to have moved Queen Victoria to tears with her rendition of "Home, Sweet Home." Born in Madrid, Spain, Adelina was the daughter of Italian opera performers. When she was young her parents emigrated to United States. At age eight she made her debut singing in concert at Tripler Hall, New York City.

Adelina Patti toured successfully in United States, Canada, Mexico, and Cuba with Norwegian violinist Ole Bull. In 1859 she made her operatic debut in *Lucia di Lammermoor* at the New York Academy of Music. She subsequently sang *La Sonnambula, Martha,* and *I Puritani.*

At her La Scala debut, Patti sang *La Traviata.* Giuseppe Verdi, its composer, declared her the greatest singer he had ever heard.

Adelina Patti toured the gold and silver mining towns, receiving roars of approval and sounds of gold pieces falling on the stage floors. Less riotous, but equally significant, were her tours of England and Europe.

In the course of her career Patti received a record $5000 per concert. She usually ended the musical evening with "Home, Sweet Home."

Patti continued to sing in concert until her last performance at the Royal Albert Hall in London, October 1924, a benefit for the Red Cross War Fund; George V and Queen Mary were among those present.

Adelina Patti died at her estate in Brecknock, Wales on September 27, 1919.

Clara Morris

Clara Morris (1847 - 1925), was known for her instinctive genius for portraying the impassioned, and often suffering, heroines of French melodrama. She was born March 17, 1847 in Toronto.

At an early age she became a ballet girl in the resident company of the Cleveland Academy of Music. After nine years of training Clara played theaters in Cincinnati, Nova Scotia, and New York City. She made such an impression on a New York director that he cast her in several highly emotional roles. Those included *Jezebel, Alixe, Camille,* and *Miss Multon.* Her most popular role was *Jane Eyre.*

Clara Morris

Clara Morris toured extensively, especially in the 1880s, and everywhere mesmerized audiences with her emotional power - a dramatic vogue of the time.

After retirement in the 1890s, Clara Morris wrote a daily newspaper column, and published several books, including *A Silent Singer*, *Little Jim Crow*, *Other Stories for Children*, *Life on the Stage*, *The Trouble Woman*, and *Dressing-Room Receptions*.

In 1904 she returned to the stage in the revival of *The Two Orphans*, and later appeared in vaudeville.

Clara Morris died in Connecticut on November 20,1925.

Adah Isaacs Menken

Adah Isaacs Menken (1835 - 1868), is best remembered by a poster depicting the heroine of *Mazeppa* bound, almost nude, to the back of a fiery coal black stallion.

"The Menken" (as she was called) was a great favorite of the miners because of her robust performance in *Mazeppa*, a nineteenth century melodrama of Gothic proportions. Wells Fargo provided a special express service to just transport Menken's trained equine performers throughout the West.

Adah Isaacs Menken was born in New Orleans in 1835. She was fluent in several languages, wrote poetry, and displayed an early talent for singing and dancing. She married Alexander Menken in Livingston, Texas in April 1856 and retained his name on stage through several marriages - all short-lived. Menken made her stage debut in New Orleans in 1857 in *Fazio*. Two years later she appeared on the New York stage in *The Soldiers Daughter*. Included in her literary circle were writers Bret Harte, Mark Twain, Henry Wadsworth Longfellow and Charles Dickens.

Menken drew crowds where ever she performed, and made several trips to Europe, delighting record audiences. In Paris she appeared in *Les Pirates de la Savane*. There she attracted the friendship of French novelists Theophile Gautier and Alexander Dumas.

On one occasion, as a token their esteem, the miners of Virginia City presented Menken with a bar of silver bullion worth $2000. Wells Fargo's strongest stage had to be specially braced to ferry it down Geiger Grade to the railroad at Reno.

On August 10, 1868, a month after Menken had collapsed during a rehearsal for a revised version of *Les Pirates,* she died.

Adah Isaacs Menken

Lotta Crabtree

Lotta Crabtree (1847 - 1924), had already attracted gold miners' attention while performing in the Califorinia mining camps. Her father had followed the Gold Rush to Grass Valley, California where Lotta met Lola Montez. The famous courtesan liked the child and gave her a bit of dancing and stagecraft lessons.

Lotta first performed before the miners when she was only seven years old. Her lively, diminutive figure immediately captivated the audience, and wither her mother, she began touring the pioneer camps.

Lotta Crabtree

In 1859 she appeared in variety theaters in and around San Francisco. Her singing and dancing kept her in the limelight. After three years of touring the country she returned to Wallack's Theatre in *Little Nell* and *The Marchioness*, adapted for her from Dickens's *The Old Curiosity Shop*, and was a sensation. From then until her retirement Lotta Crabtree enjoyed huge popularity at home and in England.

Lotta's gaiety, perennially winsome, childlike appearance, and occasionally saucy and sweetly risque manner made her a versatile performer and ranked her in the forefront of the emerging theater of burlesque extravaganza.

Her most memorable roles were were *Firefly, Hearts Ease, Topsy, The Little Detective, Musette* and *Zip.*

Lotta Crabtree died in Boston, September 25, 1924, leaving an estate of $4 million.

Helena Modjeska

Helena Modjeska, (1840 -1909), was born in Krakow, Poland. She made her stage debut in 1861 at a charity fair in Austria. Her performances in classic Polish plays as well as those of Shakespeare, Schiller, Goethe, and Moliere, soon spread her reputation. After her marriage to Karol Bozenta Chlapowski, she moved to Warsaw, becoming the leading actress at the Imperial Theatre of the city.

In 1876 she and her husband, and six others, sailed for America. They settled in Anaheim, California. Helena appeared in San Francisco, New York City and crossed the Atlantic several times for other performances.

Among her most notable roles were those in *Najezda, Odette, Daniella,* and *Mazeppa.*

One of her most memorable performances was with Edwin Booth, during which the two gave classic interpretations of Shakespeare. Booth and Modjeska brought these classics to Virginia City - favorites of Comstock- theater enthusiasts.

Helena Modjeska's grace, beauty, intelligence, and utter devotion to perfecting her art created for her a unique position on the American stage. She could spellbind an audience in virtually any role, and possessed some 260 in her repertory.

In May of 1905 a grand testimonial was arranged for Helena Modjeska at the Metropolitan Opera House by her longtime friend Ignace Paderewski.

She retired to Newport, California, where she died on April 8, 1909.

Helena Modjeska

Maguire Sells Out!

In 1875 Tom Maguire sold the Opera House to John Piper. Within a year fire wiped out his investment. Piper rebuilt, but again in 1883 fire destroyed the section of town where the Opera was located. Again the indomitable Piper rebuilt on B street where his theater stands to this day. David Belasco, one time stage manger, moved on to greater things in New York and London. Piper's Opera House gained world recognition for performances by May Robson, Lillian Russell, and Maude Adams.

Lillian Russell

Lillian Russell (1861 - 1922), was a favorite singer and actress of the pioneers. She was born December 4, 1861 in Clinton, Iowa. Helen Louise Leonard was raised in convent and private schools. About 1877 her mother took her to New York. There she had one year of opera study. Her first stage role was in the chorus of *H.M.S. Pinafore* in 1879.

Lillian Russell

Helen Louis made her New York City debut under the name of Lillian Russell, and then toured California as the lead in *Babes in the Wood.* Stardom came as D'Jemma in *The Snake Charmer*, followed by Gilbert and Sullivan's *Patience* and *The Sorcerer.*

Often in the news for her flamboyant personal life, Lillian Russell had three marriages which all ended in divorce. For 40 years she was a frequent companion of New York City's Diamond Jim Brady.

At one time Lillian Russell sought a court injunction to prevent any theater manager from requiring her to appear in silk tights. Celebrated for her beauty and her clear, pleasant soprano voice, she was dubbed "airy, fairy Lillian."

She appeared in such pieces as *The Princess Nicotine, The Goddess of Truth, An American Beauty,* and *The Wedding Day.*

After 1912 she wrote articles on beauty for the women's pages of the Chicago Herald and the Chicago Tribune. During World War I Russell was active in the Red Cross and Liberty Loan campaigns. In 1922 she toured Europe as a special investigator on immigration at the appointment of President Warren G. Harding. She died in Pittsburgh, June 6, 1922.

Maude Adams

Maude Adams (1872 - 1953), was born November 11, 1872, in Salt Lake City. She began her theatrical career at the age of nine months, when she was carried on stage by her mother in a production of *The Lost Child*. As soon as she could talk she began playing actual roles. At five she attracted attention in San Francisco in *Fritz*. At sixteen she joined Edward H. Sothern's company in New York City, making her debut in *The Paymaster*.

Later Adams played opposite John Drew. In 1897 she won top billing in *The Little Minister* . Adams played a wide variety of roles, but it was her portrayals of James M. Barrie characters that brought her acclaim, and lasting devotion of audiences.

Adams greatest successes were in *Quality Street, What Every Woman Knows, A Kiss for Cinderella,* and *Peter Pan*, in which she starred in more than 1500 performances.

Adams died in Tannersville, New York on July 17, 1953.

Maude Adams

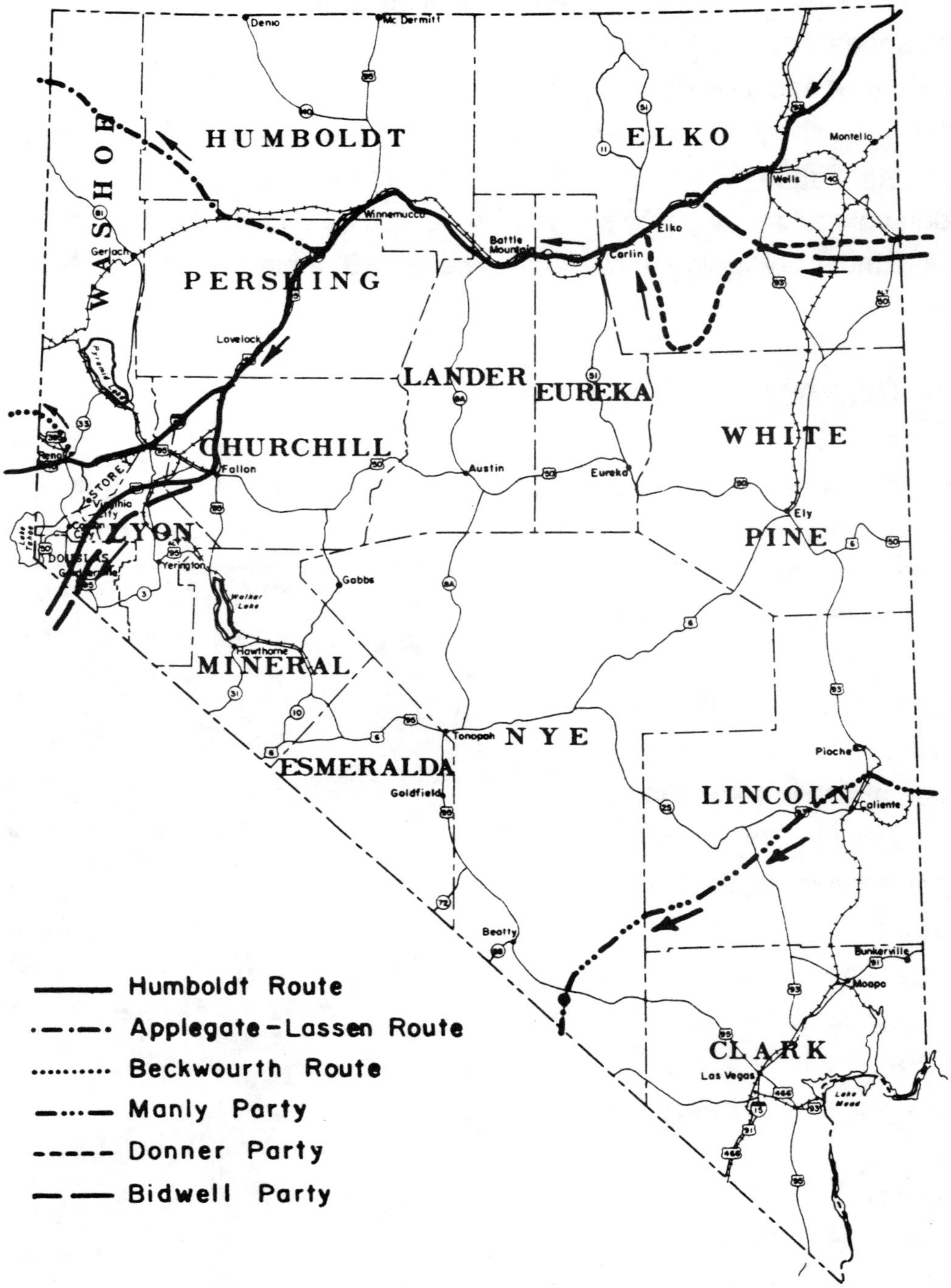

Tracing the Emigrant Routes Through the Nevada Counties and Beyond

The Sutro Tunnel

Courtesy Nevada State Museum

The Sutro Tunnel was an monumental project to tap the mines at an average depth of 1,600 feet below the surface. First favored by all the mining companies, it eventually was regarded as the bété noir, the death's head and cross-bones of every vested interest in the county. Adolph Sutro, however, proved himself no mean competitor with any who entered the drama, fighting his way inch by inch and stubbornly holding his ground with every Comstock miner, politician and capitalists of Europe.

Sutro and His Tunnel

Adolph Sutro, born 1830 in Aachen, Germany, came to United States in 1850. One of the most amazing engineering projects of the era was the Sutro Process."

Appearing in the Alta California newspaper, April 20, 1860, Sutro advocated building a tunnel from near Dayton River westward to cut the Comstock Lode at depth. To augment this proposal he would build a steam-powered reducing mill along the Carson River at Dayton, and establish a town called Sutro.

What did this mean to the Comstock miners? As they dug deeper and deeper for the precious ores, hot and cold water began flooding the underground mines. Often miners narrowly escaped drowning. Ventilation was a severe problem.

The idea of a long tunnel drew much editorial commentary. Most of the reduction mills were built along the Carson River and ore and freight, hauled by wagon and teamster was costly. The tunnel would lower the cost of transporting the ore to the Carson River, and act as a drainage tunnel too.

Sutro was encouraged with his plan by the mining companies, especially when he proposed to raise the money for the construction as well. Sutro also was smart enough to know that who ever built and owned the tunnel could do very, very well. That person could control drainage and ventilation of the mines, transport and milling of the ore, and the shipment of freight.

October 19, 1869 - Sutro strikes the first blow

Sutro mapped out the tunnel location. In late1864 he petitioned the Nevada State Legislature, and the following February was granted exclusive franchise to construct and operate a tunnel for a period of 50 years. A stipulation was that he would begin within one year and complete the tunnel in eight years.

In 1865 he organized the Sutro Tunnel Company with William M. Stewart, a U.S. Senator from the new state of Nevada, as president. Their two goals: to raise funds and to build the tunnel and to obtain contracts with the mining companies.

The mining companies would be paid two dollars for every ton of ore extracted after the tunnel reached each mine. This fee was for drainage and ventilation only. Additional fees of 25 cents per mile per ton were for removal of ores through the tunnel, and for hauling other material from the mouth of the tunnel to the mines. Transportation of men through the tunnel would cost 25 cents each. By April of 1866, 23 mining companies had signed contracts.

Graphic Promotion

To raise additional funds, Sutro hired Baron Richthofen, a well-known geologist. He prepared a handsome 83-page booklet endorsing the tunnel. With this in hand Sutro approached the U.S. government, seeking congressional consent to begin the project.

Senator Stewart also exerted influence and on July 25, 1866, President Andrew Johnson signed the consent. Sutro gained the right-of-way for a tunnel, and to purchase 4,375 acres of land at the mouth of the tunnel for $1.25 an acre. It also granted rights to any mineral lands within 2,000 feet on either side of the tunnel at $5.00 an acre. This Act superseded the Nevada Act and its time limits.

Sutro then went to New York to sell stock, claiming that the $2 million cost to dig the tunnel would yield an annual payback of between $2.3 and $6 million. He also stated that the population of the area would shift to a new town of Sutro at the mouth of the tunnel. One year later he had raised $600,000 in stock subscriptions.

In the mean time the Bank of California, controlled by William Ralston, opened a branch in Virginia City. Managed by William Sharon, the branch made loans to Comstock mining companies at a lower interest rate charged by other banks. Mining companies that soon fell on hard times, were foreclosed. Title was then transferred to the Union Mill and Mining Company, owned by the Bank of California.

By 1867 Ralston and Sharon believed that Sutro was out to undermine their power base. Comstock mining companies convinced of this threat, sided with Ralston and Sharon and withdrew their support of the Sutro tunnel.

For the next two years Sutro traveled the country trying to raise funds for his project. He wrote and published, "The Mineral Resources of the United States," describing the benefits of the tunnel. Sharon and Ralston put their own pressure on Congress, encouraging then not to get involved with "Sutro's Coyote Hole."

During these years the production of silver and gold fluctuated between good times and slow times. Sharon began thinking about a railroad from Virginia City to Carson City, and to contract with the Central Pacific Railroad along the Truckee River. To pursue this plan and destroy Sutro, he actively went after the railroad.

But on April 7, 1869 disaster struck. Fire in the Yellow Jacket mine killed 45 men, engulfing nearby Crown Point and Kentuck mines. Most of the miners died because they were unable to escape. Sutro's tunnel would have saved them!

Grasping on to the fire's devastating results, Sutro turned to the miners for support, then Washington. He lobbied President Grant but was unable to receive any government funding.

Although Sutro had not received federal support, there was sufficient interest to send out commissioners to study the tunnel project.

Meanwhile an international group of bankers advanced Sutro enough money to keep him afloat. As outside money came in Sutro began to dig. In December 1871, four shafts were started. By early 1872 the tunnel was in 2,792 feet.

About this time a report issued by the investigating Federal commissioners concluded that the tunnel was not necessary. Congress overruled their conclusions, and called for a $2 million loan to the Sutro Tunnel Company.

Work on the tunnel continued and Sutro purchased land for his town site near the tunnel portal.

Two years later, Ralston, Sharon, and their group, referred to as the "Bank Crowd," stepped up their efforts to halt the tunnel project. Now a millionaire and a senator, Sharon bought The Territorial Enterprise, Virginia Chronicle and the Gold Hills News. By controlling all three newspaper, he could control all its editorial favoritism and negativism. He quickly attacked Sutro and the tunnel

In August of 1875 the stock market took an erratic drop. The Bank of California was besieged with depositors and was forced to close.

A new group, known as the "Silver Kings,' challenged Sharon and Ralston. Ironically fate stepped in. The next day Ralston went for a swim in San Francisco Bay, and his body washed up on the shore. The "Bank Crowd" was no more.

Despite all the bank maneuvering, and the fire of October 1875, that burned most of Virginia City to the ground, the tunnel continued to advance.

The tunnel was completed June 30, 1879, with a length of almost four miles, striking the central part of the Lode at the 1600 foot level. It was 16 feet wide, 12 feet high, and contained a main drain trench in the center of the tunnel floor covered by two tracks for horse drawn ore cars.

It had cost between 3.5 and 4.5 million dollars to build. Still the many delays incurred allowed time to take its toll. By the time the tunnel was finally completed, silver had begun to play out.

Sutro ran five times, unsuccessfully, for the United States Senate. In 1880 he sold his company stock for $700,000, and moved to San Francisco. He invested in real estate, and was elected mayor of the city in 1894. He died there in 1898.

Sarah Winnemucca: Champion of the Paiutes

Sarah Winnemucca

In a time when the settlers feared the Paiutes, and Indians were treated as nomads, eking out a meager existence in meadows that once feed and clothed them, Sarah Winnemucca, stood above all others.

Granddaughter of old Chief Truckee, Sarah traveled between two societies - champion for her people, and negotiator and scout for the United States Army.

Sarah was born about 1844, a time when members of her tribe feared the guns of white men, and shunned their presence. But her grandfather welcomed the newcomers as long-lost brothers whose coming was foretold in a Paiute myth.

Chief Truckee was impressed with the white man's wonders, and when Sarah was about twelve he arranged for she and her sister, Emma, to live with a settler's family in Nevada.

It was here in Major William and Mrs. Margaret Ormsby's home Sarah became proficient in English. A pretty girl with light skin, Sarah wore long calico dresses and fashioned her hair like the white girls.

Sarah's life changed with the discovery of the Comstock Lode in 1859. Thousands of fortune seekers swamped the Paiute country, ending the tribes way of life.

The once proud Paiutes were reduced to working at menial jobs or begging for food in the mining camps. In 1860 they rose against the white man, only to be quashed and banished to a reservation at Pyramid Lake.

The army arranged for free distribution of food and clothing. But the Indian agent appointed to oversee the welfare of the Paiutes was corrupt. He sold the goods to outsiders, forcing the Indians to pay for their food. The settlers were not much better, encroaching on the reservation land.

Hostility between the Paiutes and the whites inevitably resulted into bloody encounters. One of the worst incidents occurred in 1865. Troops looking for Indian cattle thieves camp upon an Indian fishing party. They killed all of the men, women and children, including Sarah's baby brother.

During these troubling years, Chief Truckee died and Sarah's father, became chief of the tribe. Gradually Chief Winnemucca began to rely on his daughter's ability to speak English, and her counsel to ease relations between the Indians and the whites.

Sarah accompanied her father to visit the governor of Nevada, and then to San Francisco asking military officials to stop sending troops against the Paiutes.

Neither visit brought relief. But Sarah's confidence in herself and her cause was growing. For the next decade the Indians were moved from one reservation to another. In 1875 they found refuge at Oregon's Malheur Indian reservation. There Samuel Parrish, a friendly Indian agent gave Sarah a job as interpreter. The Indian received provisions and were taught to farm. Sarah, and Parrish's sister-in-law held school for the Paiute children. Unfortunately this hopeful interlude did not last. Parrish was replaced by a corrupt and self-serving agent. When Sarah attempted to complain about his conduct, he fired her and ordered her off the reservation.

Tenacious Sarah then decided to go to Washington D.C. and state her grievances. Before she got started she heard the the Bannock Indians of Idaho Territory had gone on the warpath. Among them was a captive band of Paiutes, including her father. Sarah volunteered to act as a spy for the army and infiltrate the camp.

Disguising herself in a blanket and war paint, she slipped into the camp. While the Bannocks were relaxing, she, her father and several hundred Paiutes escaped.

Sarah's daring feat brought her a reward of $500, and a job as personal interpreter and guide to General Oliver O. Howard. When the six-week campaign ended, Sarah was a heroine, and the darling of the newspapers.

Indian events, however, had a way of going wrong. The Paiutes had been promised return to the Malheur Reservation. Instead they were sent to Yakima in Washington Territory. The tribe now blamed Sarah.

Again Sarah went to San Francisco to plead the Indian case. She gave lectures, describing the self-serving and mismanaged ways of the Indian bureau and its agents. She was asked to go to Washington D.C., and to her surprise she met Interior Secretary Carl Schurz. He gave her a signed executive order authorizing the Paiutes' release from Yakima, plus a grant to each adult male of 160 acres of land at Malheur. Sarah even had a quick meeting with President Rutherford B. Hayes.

Back in Yakima the presiding Indian agent refused to recognize the government order. Sarah wrote to the President and the Secretary of the Interior but never received a reply. Humiliated and disgusted, she withdrew from public life.

Some years later Sarah again began giving lectures. This time in Boston. In 1883 she wrote, *Life Among the Paiutes: Their Wrongs and Claims*, a vivid and honest account of her life. Her friend an editor Mary Peabody Mann called it "an heroic act."

With the assistance of Boston sponsors a petition was circulated demanding that the government grant the Paiutes land. Even though it was eventually passed in congress, nothing was done to right the injustices committed against the Paiutes.

Disillusioned after repeated betrayals, Sarah permanently retired from public life, thinking herself a failure. In ill health, she died in 1891, not yet 50.

Major William Ormbsy and the Indian War of 1860

(Remington)

Major William M. Ormbsy and his wife, Margaret, had welcomed Sarah Winnemucca in to their Genoa, Nevada home. It was here Mrs. Ormbsy taught Sarah to read, write and speak English.

The Ormbsys were hospitable to the Indian girl, but Sarah long remembered the cold, military side of Major Ormsby. In September of 1857 two white men were murdered. Arrows had been found in their wounds. Major Ormbsy sent for the Sarah's brother Natchez, the peace chief, and her cousin Numaga, the war chief arrived.

"Do you know what tribe these arrows belong to?" Major Ormbsy demanded.

Numaga answered, "The Washoes."

Although the Paiutes and Washoes were not friendly, Numaga was told to bring in the Washoe leader.

Captain Jim of the Washoe tribe arrived soon after with three frightened young men, their mothers, and one young wife. Although Captain Jim knew these men did not kill the

white men, he also knew that if he didn't bring in some one, there would be war, and the Washoe's would be whipped out. Major Ormsby promised Captain Jim the men would not be hurt.

When the prisoners arrived a crowd had gathered, jeering and throwing rocks at the Indians. Through an interpreter Captain Jim was told that the men would be taken to California and tried. The terrified prisoners had little faith in the white man's justice. They bolted and tried to run away. Instantly the settlers raised their guns and fired, instantly killing two, the third Indian dying soon after.

Sarah ran to Mrs. Ormbsy, who now drew away. "My husband knows what he is doing."

Several weeks later, Major Ormbsy learned that the Indians had not killed the white men. Rather, their own men had killed them, planting the arrows so it would look like the Indians had committed the crime.

In May 1860 a war counsel was held, chief after chief rose and recounted the wrongs of his band, demanding war against the whites. Numaga spoke eloquently telling the chiefs that the white men "will come like the sand in a whirlwind and drive you from you homes. . . where you will see the women and old men starve, and listen to the cries of your children for food."

Numaga's last desperate effort to avert war was interrupted by a new-comer into the counsel.

Breathlessly he exclaimed, "Last night, with nine braves, burned Williams' station, on the Carson River, and killed four whites."

Numaga was saddened for he knew the out come. "There is no longer any use for counsel. We must prepare for war, for the soldiers will now come here to fight us."

The news of the massacre soon was carried from town to town. On Sunday, before the massacre, eight men had been hunting at Pyramid Lake when they were attacked by Indians. Five had escaped, and three were killed.

Scattered over the whole country were small groups of prospectors and ranchers, whose isolated positions rendered them as easy target to maraudering bands of Indians. In the entire area there was but one call - swift and bloody retaliation.

Gathering strength an avenging force of 105 men prepared to fight. They would include: the Genoa and Carson City Rangers, Silver City Guards, and from Virginia City were Company No. 1. and Company No. 2.

Major Ormsby was leader of the detachment from Carson City; Richard Watkins was in charge of Silver City; the Genoa squad was under the command of Thomas F. Condon, Jr.; and Archie McDonald was Captain of those men from Virginia City.

They were reportedly an undisciplined group, without a leader, but boys and men who would make a heroic defense if the need arose. It is now regarded that Major Ormbsy was the commander.

The army came down the Truckee River out of the timber, almost to the Pyramid Lake, where Namaga had taken command as War Chief. First Numaga offered a truce. The spirited band of settlers and soldiers refused.

The company mounted, and the order was given to charge. As quickly as they had appeared, the Paiutes had vanished. Quietly and cunningly Chief Namaga was luring the army down the steep river banks into a trap.

Then as suddenly as the Indians had disappeared another band, out of rifle range, appeared. The army moved forward, as if they had charged through an open gate into an Indian corral. For a time it was doubtful whether the position of Ormbsy's part was the result of accident or design; but the uncertainty vanished as every sage bush in front and on their flanks, developed the hiding place of a Paiute. A rain of bullets and wild yells followed, and in short order the battle was lost.

The retreating army headed for the timber near the plateau. This was another mistake. For there were more Indians awaiting. But even as the whites retreated, Namaga rushed ahead, waving back his warriors, trying to obtain a parley. It was too late for the Indians who, for so long, had suppressed so much rage. Yelling, they dashed by him, pursuing the army, and slaughtering most of it, including Major Ormsby. The final count stood at 84 dead. The Indian loss was acknowledged at three warriors wounded, and two horses killed.

This was the worst of all Indian wars in Nevada.

Captain Edward F. Story Leads the Retaliation

All that night refugees from the Paiute battle struggled up Sun Mountain. Outnumbered and outmaneuvered, they recounted the sudden and ghastly ambush by Numaga and his warriors.

The Comstock was in panic! Next day urgent messages for help poured over the "grape line" telegraph into Downieville, Nevada City and San Francisco. "The Paiutes were coming back!" The settlers were without protection.

A town meeting was called. Martial law declared. For their own safety, the only two women, Julia Bulette and Eilley Orrum were told to leave. Miners were warned of "jumping" claims during the emergency

After the unfortunate death of Major Ormbsy, Captain Edward F. Storey, a veteran of the Mexican War, vowed to give the Paiutes a taste of their own medicine. He organized the Virginia Riflemen.

Storey was born July 1, 1828 in Virginia, the son of Col. John Storey. In 1846, he, his two brothers and father enlisted in the Texas Rangers. Storey had married Adelia Calhoun Johnson of Texas, by whom he had one child. Six years later Adelia died. Storey and his little daughter came to Nevada. Prior to the time of the Paiute encounter, Story was engaged in stock-raising.

Storey's recruited volunteers drilled. But they had no ammunition. Story melted down silver and lead pipes into bullet molds. A fort was hastily constructed above Devil's Gate. One gifted miner built a home-made cannon out of a hollow pine log, hooped with iron hands, and mounted on a caisson. Lacking cannon ammo it was filled with scraps of iron and pieces of chain.

San Francisco responded they were sending arms and ammunition. A few days later the Sierra Battalion from Sierra City, California arrived.

On May 24, complete with martial music, fluttering flags, and drum beating, a large contingent of volunteers and regulars, marched against the Paiutes. In all there were Colonel Jack Hayes' 750 men; an advanced horse brigade of 250 horsemen; the United States Infantry from the San Francisco Presidio; and the Sierra Guards. Captain Fairfax brought up the rear were with the howitzers; followed by a commissary of beef, cattle; and pack-animals loaded down with ammunition.

With anger and retaliation burning in their hearts, the rag-tag army met the Indians at Pinnacle Mount, driving them into the waste land.

In the melee of battle, Captain Storey was ambushed, killed with a poison arrow in his lungs. He died that evening. On June 10 his body was returned to Virginia City.

He was accorded a fine funeral with all the pump and circumstance that the towns people could muster. It was led by his riderless horse with empty saddle, and boots in reverse. Next came the Masons in blue-satin aprons, their captain wearing a plumed headdress. The catafalque was drawn by six horses. Captain Storey was buried in Cemetery Ridge.

Aftermath to Battle

Numaga's skill, and his warriors, had been no match for the strong new force. He had managed to protect the women and children as they escaped, but he lost the battle, and with it the war.

With the fighting over, and Numaga still serving as chief, Winnemucca was no longer needed in Nevada. He took his family to join the distant Paiute tribes in Oregon.

Sarah Winnemucca probably fled with the tribe. Ironically neither side wanted to continue the fight. The settlers wanted an end to the Indian attacks, and the Indians were tired of hiding in the rocks, and were desperately hungry.

Numaga met with the army officers, and bitterly told them the white men were "like coyotes, . . . making a heap of talk and much of their talk was not good."

Nevertheless Numaga pledged the Paiutes would not make further war for a year, provided the white men also keep the peace.

Sarah and Winnemucca were called back. She acted as translator. With him, she met with the army, promising them to lay down arms in return for protection, and a reservation at Pyramid and Muddy lakes. The army acknowledged their requests. They would give the Paiutes food and supplies, teachers, and aid in learning to farm. For now life appeared peaceful for the Paiutes and the settlers.

Old Chief Poito Winnemucca, whose father Chief Truckee had welcomed the white men as long lost brothers, whose coming was foretold in an old Paiute myth. Winnemucca was the father of Sarah Winnemucca, and the uncle of the war chief Numaga.

In1850 a trading post was established in northwestern Nevada named French Ford. With the coming of the Central Pacific Railroad in 1868 it became a terminal and major shipping point. In honor of the old Paiute Chief, it was renamed Winnemucca.

Romans One - Miners Two!

The Mormons had denounced Virginia City, saying they didn't want any part of it because of the excessive drinking and other nefarious activities. To some betting was also work of the devil. To the hard working, hard drinking lusty miners it was just another form of their wild entertainment. They were not unlike the Romans with their insidious amusements.

Rollicking, hair-raising events such as horse races, cock fights, dog fights, badger contests and prize fights were eagerly anticipated. To avoid the law, (for most of these events were prohibited), the games were often held out of town.

As one writer noted, "Had it been possible to hang men and afterwards resuscitate them, or blow them from the mouths of cannons, and then gather their fragments together and breathe life into them - such extravagant feats would have sufficed."

Prize fights with raw, bare fists, were generally held in out-of-the-way places. They commenced as early as 1863 when a certain pugilist named Bradley and and a big Irishman named Pat Fogg arranged a fist fight for money.

The following January, in honor of General Jackson, (a sensible reason the miners concluded) a fist fight was arranged in a vacant lot on B Street. But to the disgust of many spectators it was halted by police.

In March, Fogg and Bradley met again, this time in Six-mile Canyon. There is no indication who came off best.

Bill Davis and Pat Dayley had a go- round at American Flat in June for $1000 a side. Three thousand spectators witnessed the event, paying $2.50 a head. Again there was no account of the winner.

Fogg entered the arena again to fight William Blackwood. The event was held at Spring Valley. After twenty-four rounds the victory was awarded to Fogg.

Fogg never tired of fighting. He put up his fists the following year. That time he went one hour and twenty-two minutes with Harry Cooper. At the close of the sixty-third round Fogg was declared winner in a "prolonged and brutal affair."

In the next fight Fogg whipped John Kelleher. They went thirty rounds, in one hour and seventeen minutes. Stakes were $1000. The match was held on the border of Storey County.

The last of the recorded exhibitions occurred in 1872 on the old race course between White, an Irishman, and Williams, a Cornishman. Stakes were $250 a side. Williams won.

In 1865 a dog and bear fight came off at Maguire's Opera House. The bear made short work of whipping the dog - just minutes before the police arrested the managers.

The biggest prize fight had to have been the Heenan - Sayers bout between American heavyweight champion, John C. Heenan, and Tom Sayers of England, April 17, 1860. The championship event was held at Farnborough, England.

Nothing was more important to the miners than this fight. Heenan, known as"Benicia Boy," was prominent in Benicia, California. Miners had worked with Heenan. In Washoe there was a "Heenan Bar," a "Heenan Sample Rooms" on Sun Mountain. Heenan was even married to the darling of the Comstock, Adah Isaacs Menken.

If all went accordingly fans would hear the fight results on May 6th or 7th, depending which day the pony express arrived. On May 8th, however, word of another battle fired the miners. "Paiutes had attacked Williams' Station. . . . Arm yourselves - quick!" came the call.

The Indian war momentarily took president over the prize fight.

Still, miners found plenty to remember that year. After forty-two rounds, spectators broke into the ring. The Sayers-Heenan fight was declared a draw. Each contestant was awarded a championship belt.

In 1865 the Queensberry rules were drawn up. Their purpose was to get rid of the brutality of professional prize fighting and make boxing a sport. The first use of boxing gloves occurred in 1872.

Julia Bulette: *"The Girl I Left Behind"*

At eight o'clock on an early spring evening in 1864 Virginia City's finest and roughest were packed into Maguire's Opera House awaiting the melodramatic performance of Adah Isaacs Menken. Crystal chandeliers glittered, dancing off the gems of the town's richest clientele.

Amidst the glow and finery was Julia Bulette. She had arrived in a brougham coach with her personal crest emblazoned on its door. Great diamonds sparkled in the lobes of her ears. A ruby gulled on her bosom. Precious sables nestled about her white satin shoulders. Julia was happy - she had risen above her calling.

Julia Bulette

Julia had told the world of Sun Mountain she was French. Others said she was English. While acquaintances in New Orleans, claimed she was plain "Gulf." Whoever she was her true identity was never disclosed.

None the less the town was devoted to her. "Her skin may be scarlet," said one , "but her heart is white."

By worthy deeds, Julia seemed anxious to compensate for those by which she earned her living. For Julia was the town's most famous and richest courtesan. She was queen of D Street, a legend in her own time.

Julia and Eilley Bowers were the first women in Virginia City. Julia had cared

for the sick during an influenza epidemic, carrying medicine and comfort from shack to tent. She cheered the men on when the Paiutes threatened the inhabitants. And when silver made the prospectors crazy with wealth, it was Julia who graciously opened her home to all who could pay her price.

Julia's Palace became known not only for its male pleasures, but as a stylish dining establishment that offered fine wines and French cooking. Julia spared nothing for her adoring customers.

Julia was honored by members of the Fire Engine Company No. 1. They named a Virginia &Truckee car for her, and even a mine. The Menken may have been the most famous entertainer to descend on Virginia City, but Julia was the bravest, most courageous and compassionate.

As with many great and noble deeds, however, a violent ending was to take Julia's life. For on one wintry night, she was killed for the diamonds that adorned her ears, the ruby that coruscated on her breast, the sable scarfs that nestled about her throat. John Millain was said to have strangled her to death on her ornate canopied bed.

Of course, Millain was hanged. The town had no mercy for the murder of their finest madame.

A full municipal holiday was declared for the occasion. Millain arrived in an open barouche accompanied by Father Manogue and followed by other state and local dignitaries, all wearing top hats and frock coats. Members of The Territorial Enterprise staff were invited on the scaffold. A crowd of 5,000, all in a festive mood and

John Millain's execution - the California Police Gazette

attire, joined in the festival-like atmosphere.

According to the next day's newspaper reports, Millain was smoothly shaven, and looked somewhat pale from imprisonment, but showed no sign of nervous weakening. He stepped forward and read his farewell address in French, saying "a great injustice had been done him. He had been guilty of some bad deeds in his lifetime, but of this murder he was innocent. Then Millain thanked the ladies who brought him cake in jail, shook hands with the officers, embraced the priests, and stood firmly on the fatal trap while the noose was adjusted about his neck. The crowd roared as the trap door was sprung.

John Millain

Julia's last hurrah was nearly as celebrated. All the town's enterprises shut down and the entire city turned out for her funeral. Julia was still a member of the fire company at the time of her death. Therefore it was proper that she should be buried by the company. The men of Fire Engine Company No. 1 arrived in full dress and marched behind her silver-handled casket. It was carried in a glass-walled hearse to the Flowery Cemetery east of town.

At Julia's grave site the red-shirted firemen paid her a last tribute when they band played "The Girl I Left Behind ."

"The Girl I Left Behind "

Aid for Civil War Victims - Blue and Gray

Two thousand or so miles from the battlefields of the Civil War didn't matter to the pioneers. Soldiers from the South and the North were lying in hospitals, sick and wounded. They needed bandages, condensed milk, beef stock, spirits and coffee.

Weren't the billions of dollars springing from Sun Mountain worth sharing with them? The prospectors were in agreement.

In 1864 sympathizers of both North and South, offered their support to the boys in Blue and Gray. Money was to be raised by donation and auction and sent to the United States Sanitary Commission in New York City.

The gimmick, if you will, was bidding on a sack of flour. Seems that the sack of flour started in a municipal election in Austin, Nevada. Charles Holbrook, an ardent Unionist, ran against an equally red-hot Democrat.

David E. Buel, who considered himself a Southerner, made a wager. If his opponent, a rabid secessionist named Holbrook was elected, Buel agreed to carry a sack of flour on his back, decorated with stars and strips, through the streets of Austin.

Holbrook was elected and Reuel Gridley, a friend of Buels, offered to shoulder the sack. Gridley was a good sport. A rally was held with all the patriotic trimmings. Mark Twain dubbed the fundraiser "The Army of the Lord." When the band played both "Dixie" and "John Brown's Body," the crowd went wild with patriotism. Austin raised $6000 for the boys in Blue and Gray.

Virginia City heard about the gesture and when Gridley showed up in town, enthusiasm spilled over. The Comstock folk decorated four wagons, and engaged a brass band of fourteen.

But to everyone's surprise, "The Army of the Lord," raced off to Gold Hill. The Comstock were stunned. In Gold Hill the cavalcade was greeted by men, women, children,

Chinese, and Paiutes, and the flour sack. Bidding ended with $7052. Gold Hill had beat Austin's bid.

Virginia City had been out stripped by their great rival, Gold Hill. With tempers flaring, Gridley and the wagons were encouraged to return. By night fall, the town was ablaze with flags, torches, bands playing, crowds cheering. The wagon procession had swelled to a mile long, adding as it went, National Guards, Regulars from Fort Churchill, and a mounted cavalry. When the parade finally halted in the center of town, on-lookers were wild with excitement.

"The greatest self-rising flour in the country!" went the bidding cheers. Residents might be at odds over the issues of the Civil War, but if American soldiers needed aid, they were united in one cause.

Late that night charity prevailed. The Comstock folk had raised $275,000 on one sack of flour. The money collected was melted down into fourteen silver bricks. On each brick was stamped the name of a battle. The silver was then delivered to the Sanitary Commission in New York City.

In the spring of 1861 Wells Fargo began to bear heavily on the Pony Express. In this ad Wells Fargo appealed for business in New York under the reduced rate of $1.

Nevada Becomes a State

Nevada Territory was created on March 2, 1861. On October 31, 1864, three years later, Nevada was proclaimed the thrity-sixth state. This unprecedented brief period of time between establishment of a territory and the granting of statehood, owed much to the fabled riches of the Comstock Lode.

Politics was the major factor for Nevada's rapid acquisition of statehood status. The Republican party under, President Abraham Lincoln, wanted another GOP state, which Nevada was. The reason: to send its Congressional representatives to Washington, D.C. in support of the president's political policies during and after the Civil War.

However, while it is still debated by scholars as to the actual effect that the silver taken from the mines under Virginia City had on the Civil War, it is clear that the anticipated wealth from the Nevada mines was a factor in the politics of the state's admittance.

But before statehood could be granted, a convention was held to write a state constitution. It was hastily adopted and paved the way for statehood.

To save time in proclaiming statehood and to enable the voters to take part in the presidential campaign, the new constitution was telegraphed to Washington at a cost of $3,416.77.

In the early days of the state, many "bonanza kings" are said to have bought their elections to the United States Senate. To prevent dishonest elections, Nevada passed a corrupt-practices act. It limited campaign expenses to 20 percent of one year's salary and set up severe penalties for bribery. This law was repealed in 1951.

In 1914 an amendment allowing women to vote was passed - six years before the federal amendment.

John MacKay the "Bonanza King"

The Comstock Lode made millionaires of many persons, but there was only one who received the title of the "Bonanza King" - an Irish emigrant named John MacKay.

John MacKay

MacKay was born November 28, 1835 in Dublin, Ireland. He came to California in the 1850s to make his fortune in the gold fields. For most of a decade he toiled without ever reaching his dream. When silver was discovered on the Comstock, MacKay decided to try his luck there.

Upon arriving in Virginia City, he hired miner at wages of $4 a day. His intelligence and perseverance soon enabled him to form his own mining-contracting company. Financial success in this business allowed him the opportunity to invest in several mining partnerships during one of the Comstock's many depression periods. MacKay's wealth and reputation grew.

In 1867 MacKay married Marie Bryant, widow of a Downieville surgeon and daughter of Col. Daniel E. Hungerford, who had served in the Indian campaign of 1860.

In 1869 with James G. Fair, MacKay contracted to develop th Hale & Norcross Mine. Although the stock fell, he had faith in the project and induced James Flood and William O'Brien, successful mining operators in San Francisco, to aid in securing control. J.M. Walker also became a partner. They extended their possessions until they controlled 3,000 feet of the Comstock vein, north of the Hale & Norcross. They set an exploring drift along that property. Walker became discouraged and sold his one-fifth interest to MacKay.

The result of the drift was the discovery of the "bonanza," in the Consolidated Virginia and California Mines. Between the years 1874 - 1879, the mine paid its stockholders $100,000,000. MacKay's fortune escalated, and he was now referred to as the "Bonanza King," and a member of the group known as the "Big Four," or "Silver Kings."

MacKay, his wife and two children took full advantage of his fantastic wealth to help make Virginia City a cultural center between Denver and San Francisco. His wife spent much of her time in Paris. After the fire that destroyed Virginia City in 1875, MacKay used his money for civic improvement such as rebuilding St. Mary's Church, and built a cable across the Atlantic Ocean.

At MacKay's death in 1902, his business manager told reporters: "I don't suppose he knew within twenty millions what he was worth."

MacKay's widow and son and daughter-in-law donated several millions of dollars to construct a building in his name at the University of Nevada, Reno, for teaching of mining engineering. It evolved into the renowned Mackay School of Mines.

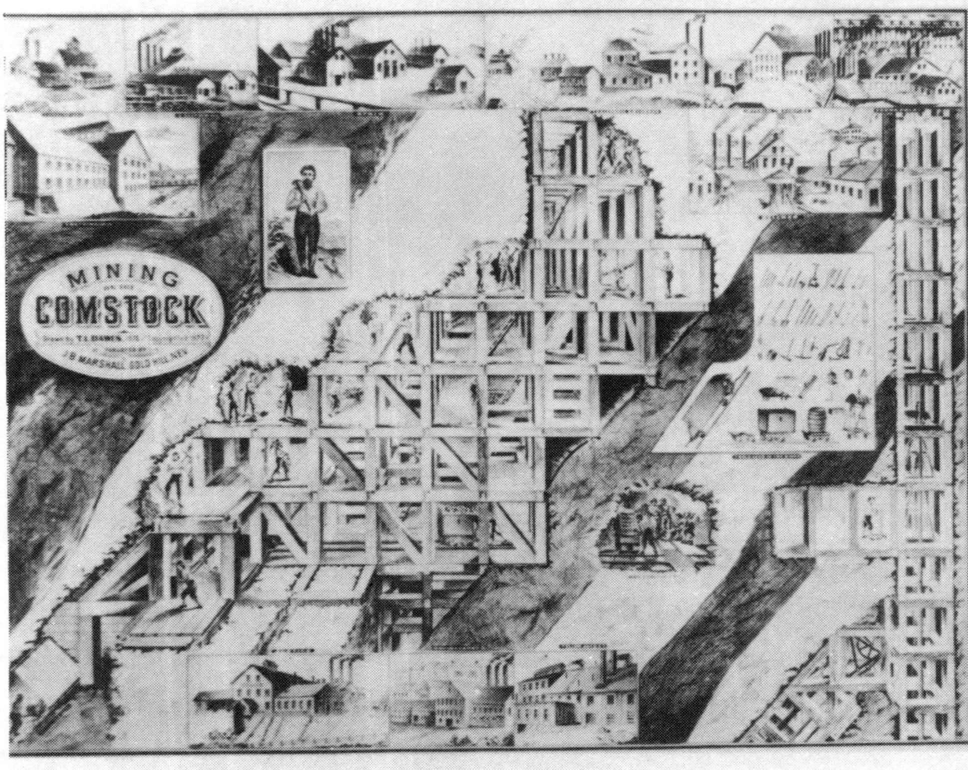

MINING *Methods*

So rich was the fabulous Comstock Lode in Virginia City that a new method of mining had to be invented to work the veins. Center is pictured the Square set timbering devised by Philip Deidesheimer in the 1860's, and now used throughout the world. Note the cooling rooms for relief from the terrific heat.

Enter the "Silver Kings"

James G. Fair

James G. Fair was born in Clougher, County Tyrone, Ireland, December 3, 1831. Twelve years later he and his parents came to America.

His first taste of mining was an an eighteen year old, placer mining along the Feather River area of California. Looking for the source of minerals soon asserted itself, and Fair moved on to the Angel's Camp area of California. Later he became superintendent of quartz mines in other parts of the state.

With the Comstock discovery Fair found work as superintendent at the Ophir Mine, and in 1867 at the Hale & Norcross Mine. There, began his association with John MacKay.

MacKay and Fair needed more capital to gain control of Hale & Norcross. They approached two old friends, James Flood and William O'Brien, and together they were able to purchase the mine. In 1870 a new body of ore was discovered. The result was the unprecedented wealth of the "Silver Kings."

With his share of profits Fair acquired large real estate holdings in San Francisco, including the Fairmont Hotel, and mining interests in Arizona and Georgia.

Despite his increasing wealth, Fair spent many days and nights visiting the different workings of the Comstock mines. He was said to be knowledgeable with "every cross cut, winzes, stopes, shafts and inclines of the mines."

Fair, with his wife and four children, later toured the world. Upon returning home he sought the office of United States Senate, then held by William Sharon. Although he was a Democrat, and the new state had previously been represented by a Republican, the people desired change. Fair was elected Senator in 1881, holding the position until 1887.

At one time Fair's net worth was estimated at between forty and firty million. Even with Fair's great wealth, he was acknowledged as "having been placed in charge of a great property, and there he faithfully remained, toiling as when years before necessity required him to carefully husband his daily salary."

James C. Flood and William O'Brien

James C. Flood, was also a partner of the "Silver Kings." He was born in New York City in 1826. He arrived in California during the Gold Rush of 1849.

Flood's early years were spent in the retail liquor business with William S. O'Brien. It was from the funds of the Auction Lunch Saloon in San Francisco that they were able to accumulated enough money to purchase portions of the now famous Hale & Norcross Mine.

John MacKay, with the help of Flood and O'Brien, secured the claim which grew into the mining bonanza. Flood's executive ability proved worthy of the challenge. With $30,000,000 thrown into his hands, Flood proved himself a first-class financier.

He built himself two grand homes, one in Menlo Park, and the other atop Nob Hill, San Francisco, both in California.

Shy and soft-spoken William O'Brien, lived out his later years in Marin County, California.

In 1872 the "Silver Kings" began looking for other properties to explore. Quietly they began purchasing stock in the Consolidated Virginia & California Mine. Because of money problems, the mine-owners (the "Bank Crowd") had only explored it to a depth of about 500 feet. The "Silver Kings" believed they would find ore at a deeper depth. They were able to take control of the mine for less than $50,000.

In 1873 the top of the "Big Bonanza" ore body was found. Within the next four years the "Big Bonanza" yielded 1.4 million tons of gold-silver ore totalling $105 million. Seventy-four million was reportedly paid in stock dividends. The "Silver Kings" were now truly kings of mighty riches.

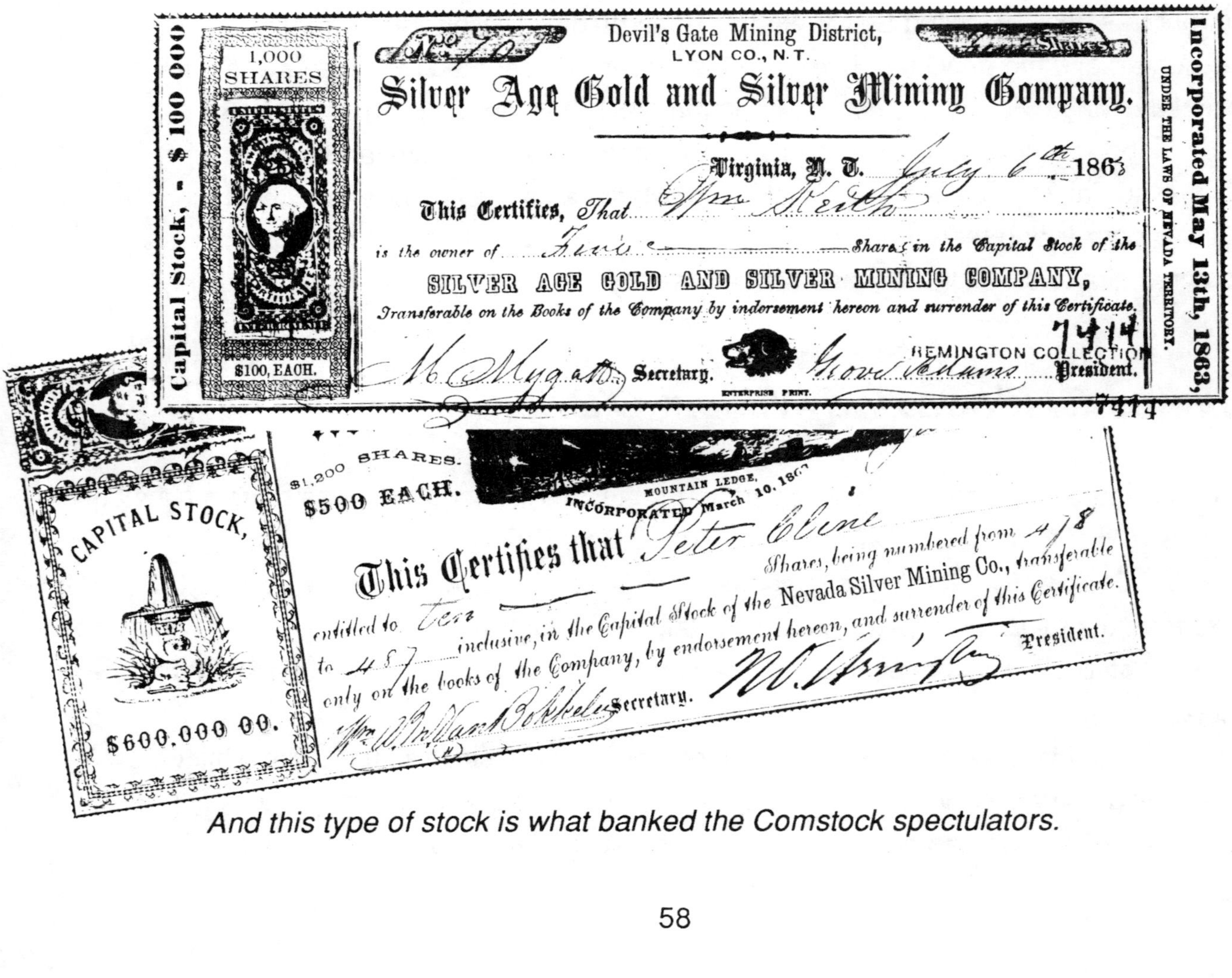

And this type of stock is what banked the Comstock spectulators.

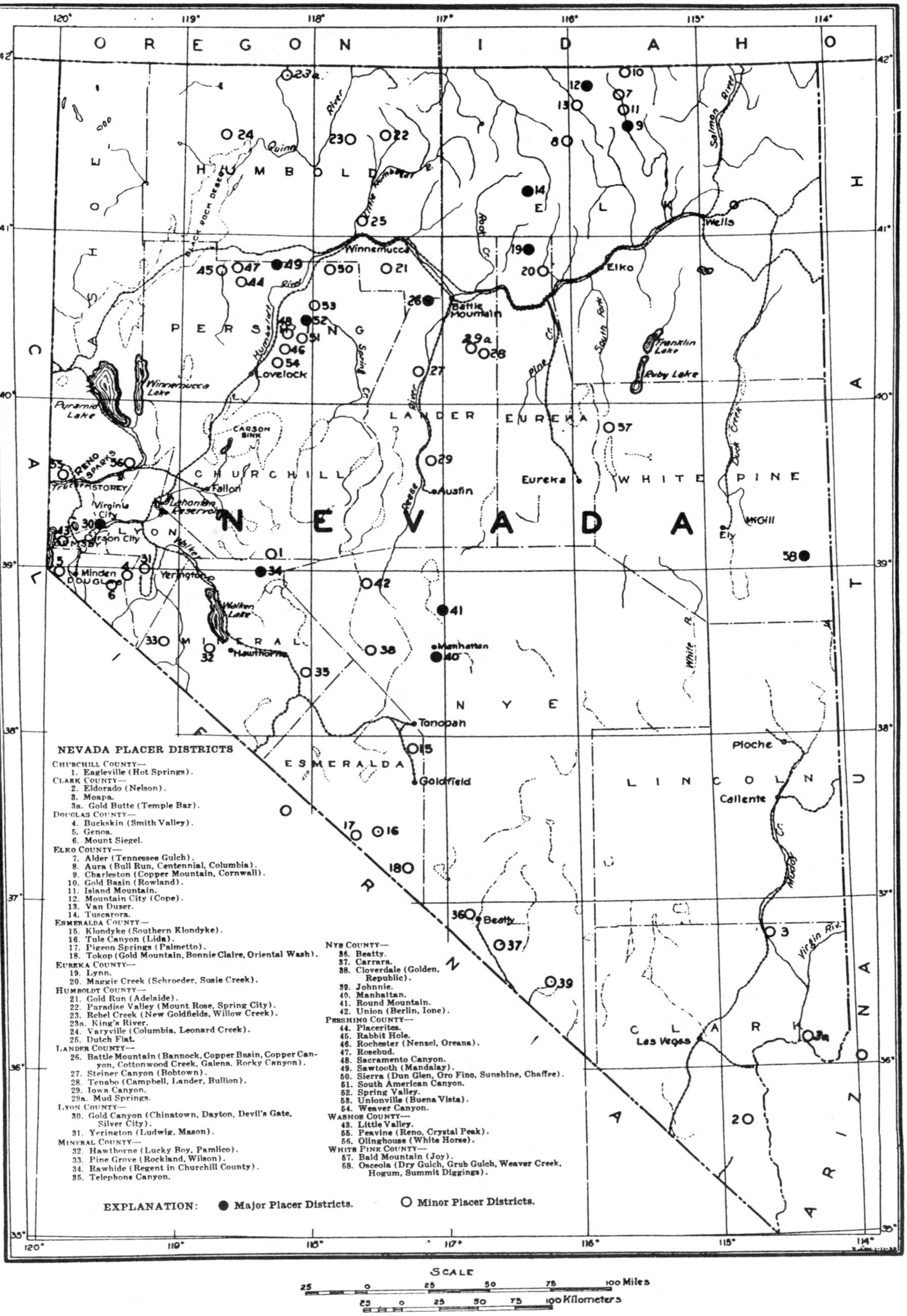
NEVADA PLACER DISTRICTS
CHURCHILL COUNTY—
1. Eagleville (Hot Springs).
CLARK COUNTY—
2. Eldorado (Nelson).
3. Moapa.
3a. Gold Butte (Temple Bar).
DOUGLAS COUNTY—
4. Buckskin (Smith Valley).
5. Genoa.
6. Mount Siegel.
ELKO COUNTY—
7. Alder (Tennessee Gulch).
8. Aura (Bull Run, Centennial, Columbia).
9. Charleston (Copper Mountain, Cornwall).
10. Gold Basin (Rowland).
11. Island Mountain.
12. Mountain City (Cope).
13. Van Duzer.
14. Tuscarora.
ESMERALDA COUNTY—
15. Klondyke (Southern Klondyke).
16. Tule Canyon (Lida).
17. Pigeon Springs (Palmetto).
18. Tokop (Gold Mountain, Bonnie Claire, Oriental Wash).
EUREKA COUNTY—
19. Lynn.
20. Maggie Creek (Schroeder, Susie Creek).
HUMBOLDT COUNTY—
21. Gold Run (Adelaide).
22. Paradise Valley (Mount Rose, Spring City).
23. Rebel Creek (New Goldfields, Willow Creek).
23a. King's River.
24. Varyville (Columbia, Leonard Creek).
25. Dutch Flat.
LANDER COUNTY—
26. Battle Mountain (Bannock, Copper Basin, Copper Canyon, Cottonwood Creek, Galena, Rocky Canyon).
27. Steiner Canyon (Bobtown).
28. Tenabo (Campbell, Lander, Bullion).
29. Iowa Canyon.
29a. Mud Springs.
LYON COUNTY—
30. Gold Canyon (Chinatown, Dayton, Devil's Gate, Silver City).
31. Yerington (Ludwig, Mason).
MINERAL COUNTY—
32. Hawthorne (Lucky Boy, Pamlico).
33. Pine Grove (Rockland, Wilson).
34. Rawhide (Regent in Churchill County).
35. Telephone Canyon.
NYE COUNTY—
36. Beatty.
37. Carrara.
38. Cloverdale (Golden, Republic).
39. Johnnie.
40. Manhattan.
41. Round Mountain.
42. Union (Berlin, Ione).
PERSHING COUNTY—
44. Placerites.
45. Rabbit Hole.
46. Rochester (Nenzel, Oreana).
47. Rosebud.
48. Sacramento Canyon.
49. Sawtooth (Mandalay).
50. Sierra (Dun Glen, Oro Fino, Sunshine, Chaffee).
51. South American Canyon.
52. Spring Valley.
53. Unionville (Buena Vista).
54. Weaver Canyon.
WASHOE COUNTY—
43. Little Valley.
55. Peavine (Reno, Crystal Peak).
56. Olinghouse (White Horse).
WHITE PINE COUNTY—
57. Bald Mountain (Joy).
58. Osceola (Dry Gulch, Grub Gulch, Weaver Creek, Hogum, Summit Diggings).
EXPLANATION: ● Major Placer Districts. ○ Minor Placer Districts.
SCALE
25 0 25 50 75 100 Miles
25 0 25 50 75 100 Kilometers

Major Virginia City Fires

The combustible nature of the materials of which Virginia City was built, the fierce winds which daily blew through the mountains, the dry character of the atmosphere, and the careless habits of the transient population all combined to make the chances of devastation by fire numerous:

August 29, 1863. The first great fire broke out in a carpenter shop in the rear of Patrick Lynch's saloon. About $700,000 worth of property was destroyed. The fire burned from Taylor Street to Sutton Avenue, north and south, and from A Street to B, and partly down C, east and west. In 1863 this was the principal business part of town.

September 29, 1865. Fire started at the Fountain Head Restaurant and burned over an acre extending from Union Street to below Sutton Avenue, and as far as D Street east, and A Street west. Estimated damage: $400,000.

September 23, 1866. The Music Hall was destroyed by a fire caused by a bursting lamp.

September 19, 1870. Fire in Lonkey & Smith's lumber yard spread from D to E Street east, and as far as A Street west. Total damage was $300,000.

June 29, 1873. At 11 p.m. the McLaughlin & Root building on B Street blew up and took fire, damaging a portion of the Bank of California.

One hundred pounds of Hercules powder, six cases of nitro-glycerine, 100 pounds of giant powder, and 200 pounds of common powder had been stored under the bedroom of Major General Van Bokkelen. In the ensuing explosion and fire he was killed, as was J.P. Smith, W.D. Davis, Ben Mandel, Mrs. Ed Dean and daughter, Billy Low, Charles Knox, Mrs. Emily Connor and John Devine. Many other persons were wounded. Loss was set at

$225,000.

October 26, 1875 - remembered as the great fire, started at 5:30 a.m. in a lodging house on A Street, taking four grueling days to be put out.

It resulted in the total destruction of the business portion of the city, at a loss of about $12,000,000. The burned area included every thing between Taylor Street on the south, and Carson Street on the north, Steward Street on the west, and the Chinese quarter on the east. The fire soon was out of control, leaping across whole blocks. The mills and hoisting

works were swept away in the fire storm.

Considerable quantities of blasting powder was stored around town, and the erratic explosions increased the horror. The fear of more explosions paralyzed the efforts to quell the fire and to remove valuables.

Mining shafts burned down to a considerable distance, causing cave-ins. At Ophir, a cage was let down filled with dirt to prevent the fire from passing down, but the fastenings gave way when the dirt was shoveled on it, and the fire roared down the shaft.

Soon after the fire a gale commenced blowing, knocking down the weaken walls and filling the air with clouds of ashes and cinders. A snow storm set in on November 2nd, making the situation worse.

In all over 300 businesses were destroyed, and 1,000 dwellings. The mill of the Consolidated Virginia mine was destroyed at a loss of $500,000.

Residents quickly began to rebuild, and shelter was provided in a few days. The

Ophir mine shaft was quickly repaired and operating in thirty days. The railroad brought in supplies of timber and provisions. Without the railroad, said residents, the city would have been abandoned until spring.

April 7, 1869. At 7 a.m. a burning candle accidentally left by a night-shift miner started a fire in the Yellow Jacket Mine. It soon engulfed the dry timbers, sending fire through the shafts. Thirty-six miners were killed.

September 19, 1873. A second fire erupted in the Yellow Jacket Mine started by sparks from a forge. Four miners died and dozens were wounded.

The Virginia City Volunteer Fire Company

The origins of the Virginia City Fire Company started as a bucket brigade. In the fall of 1860 volunteers who organized to fight fire had only buckets. The following year this procedure was disbanded and two fire departments were formed; Hook and Ladder Company No. 1, with 65 members, and Virginia Engine Company No. 1, with 26 men. Each company had an engine and 600 feet of hose.

Thereafter, on March 19, 1862 the Young American Engine Company, No. 2, was organized with one $6,000 engine, and 80 men.

Eagle Engine Company, No. 3, was organized on August 2, 1863, and purchased a hand engine for $3,700 belonging to Vigilant Company, No. 9, of San Francisco.

Washoe Engine Company, No. 4, was organized soon after with 80 members.

Knickerbocker Engine Company, No. 5, was organized in the summer of 1864.

Confidence Engine Company was organized October 19, 1864.

In May of 1866 the Board of Aldermen of Virginia City granted the Confidence Engine Company the right to change their name to Monumental, No. 6.

Hand-in-Hand Hose Company, No. 1, was organized November of 1866, but shortly thereafter disbanded.

These companies were the sum total of Virginia City's volunteer fire fighters.

Royal Times for Sandy and Eilley : Comstock's First Millionaires

Sandy Bowers was a genial and somewhat illiterate miner. Eilley Orrum washed the miners clothes, provided them board and room, and feed them home cooking.

Before the big bonanza Sandy staked out a small claim in the center of the lode, stubbornly refusing to part with it for trivial rewards. Adjacent to Sandy's ten feet were ten more feet owned by his landlady and laundress. Although Eilley had already discarded two Mormon husbands, love bloomed once again. She and Sandy were married, consolidating their ten percent into twenty percent.

Overnight their twenty percent suddenly turned to precious gold and silver. Sandy and Eilley became the first millionaires on the Comstock. Near by the Gold Hill stamp mill crushed their precious ore to a fantastic tune of $100,000 a month.

Unlike other miners who's instant wealth would send them scurrying to Nob Hill, Sandy and Eilley had a grand time with their money - well it lasted.

With thousands and thousands of dollars a month to spend, Eilley indulged in another of her favorite past times - crystal gazing. She saw herself a ruler - Queen of Washoe! The miners were thrilled when she told them

Eilley Orrum Bowers

the news. Thereafter when they met her on the trail they doffed their dusty hats and called her "Your Majesty."

Now that Eilley was called "queen," she must have a castle. The plans called for an expenditure of half-a-million dollars - pennies for Sandy and Eilley.

Marble was freighted over the Sierra. Silver door knobs and silver hinges were hammered out of bullion from their mill. Hot-water, from a near by thermal spring, gushed out of the mouths of silver-headed mountain lions. In the library morocco-bound books, all bought by the yard, graced the shelves. Lace curtains hung about glass windows. In the garden were fountains adorned with Spanish tile, statues, and stone towers. It was all so magnificent.

Friends saw the house and exclaimed it was more like a palace with all its gold gilt, lace and plush. When it was completed, the miners told Eilley she should get acquainted with the Queen of England. Eilley agreed that all queens should know one another. She would take the Queen a gift of silver, in turn ask for cuttings of English ivy from the walls of Windsor Castle to climb over her own castle towers.

Sandy and Eilley left Virginia City with a quarter-of-a-million dollars and an oak chest filled with silver bullion. Journeying over the Sierra, a Wells Fargo special armed guard accompanied them to protect the chest from bandits.

In San Francisco the silver bullion was converted into a beautifully designed dinner service - their gift to the Queen. Then the Bowers set sail.

The day royal court was held, Eilley wore three feathers, a veil, and a long court train. Some said Eilley never met Queen Victoria, (royals frowned on multiple marriages). Others said that "when she approached Her Majesty the Queen of England, Eilley made a handsome courtesy." Its nice to believe that Eilley was indeed received.

The silver service was presented, and Eilley received the sprigs of English ivy.

Sandy and Eilley made the Grand Tour of Europe, and remained for several years. Back home their claims continued to produce fantastic sums.

When the Bowers finally returned home, they brought French mirrors, Italian statuary, priceless oil paintings, crystal chandeliers, and suites of bedroom furniture.

In the late sixties Sandy died and was burned in the hillside behind his splendid home. In time their bountiful claim ran out. Eilley was forced to leave her castle, living the rest of her life in poverty, at last joining Sandy on the Washoe hillside. Their home stands today, a monument to the extraordinary wealth of the Comstock Lode.

Newspapers of the Day

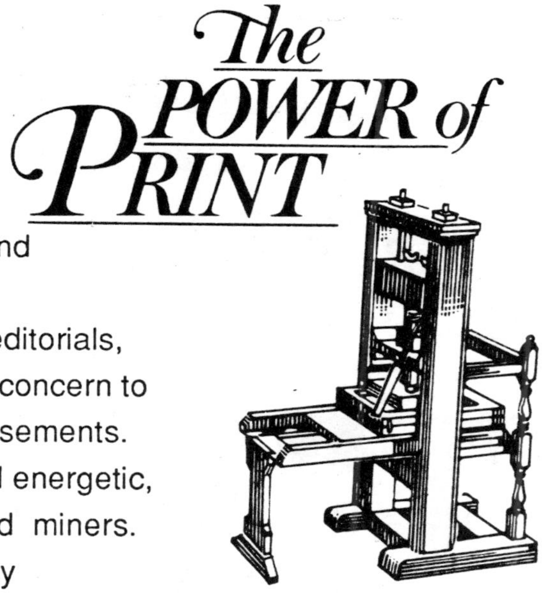

Along with loving riotous times and their whiskey, the pioneers cherished their newspapers. When the morning drink was consumed, the morning paper was read and devoured as a matter of course.

The press of1860s was known for vigorous editorials, and first class satire on town gossip. It ran items of concern to citizens and supported by subscriptions and advertisements.

Newspaper readers were both intelligent and energetic, and a large mixture of the reckless and rough-edged miners.

Newspaper content demanded editorial ability to match; firmness, honesty, and the ability to treat current questions of the day with clarity, sensibility, and a lively turn. Not withstanding were their jabs at local issues, and hoaxes, generously contributed by the writers and publishers. All of which was bent on satisfying the readers insatiable appetite for news.

News from the east coast and Washington was often garnered from eastern papers. It might be a tad late in appearing in the local broad sheets, but it always was of keen interest, and believed critical to the pioneer's way of life.

The Territory Enterprise held title as the pioneer newspaper. Under the management of Joe Goodman and Dennis McCarthy, who had been printers before becoming publishers, it became a first-rate newspaper with a hearty readership.

Mark Twain and Dan DeQuille enriched the paper with their humorous writings.

There were, however, other newspapers during the bonanza period. The Daily Union was established in November of 1862 by John Church and Company. It claimed the largest circulation. It began as a Republican newspaper, but changed its views to Democratic when T.L. Ham and J. H. Hum withdrew from the firm.

To further support the Democratic cause, there was the Daily Democratic Standard, first appearing August, 1863. With the end of the electoral campaign it closed down. At that time here were also three Republican newspapers.

The Daily Old Paiute was started in 1865 by John K. Lovejoy and Company, and was distributed as an evening paper. It's politics were Republican, and printed "with a pointed pun and vulgar turn." It eventually suspended operation.

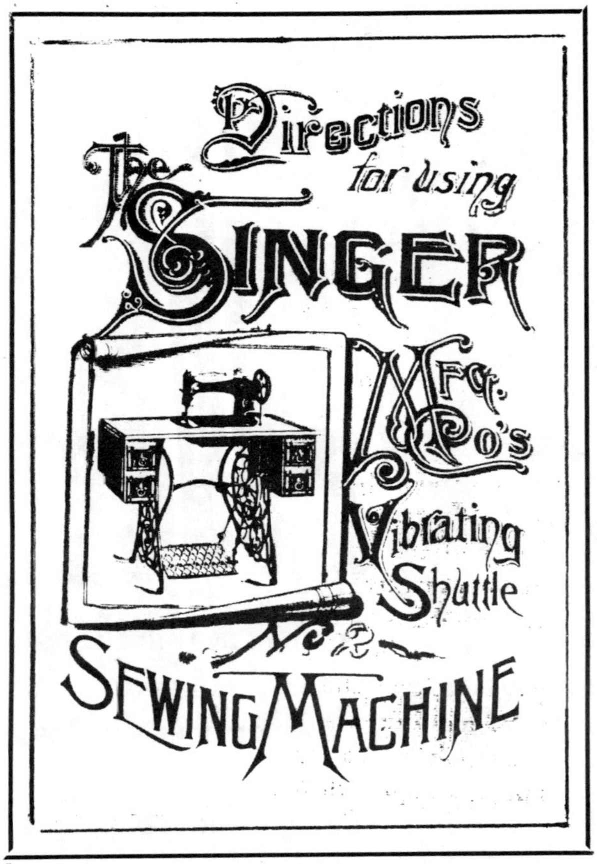

The Nevada Pioneer, was a German newspaper, published semi-weekly and Democratic in its views. It was started in March, 1864, at Virginia City, by J.F. Hahnlen. It was suspended seven months after its debut.

Philip Lynch started the Gold Hill Daily News in October, 1863. It was an important paper in the state.

The Washoe Daily Evening Herald, July, 1864, started when there was already four newspapers in Virginia City. No matter, it was Republican in principals, with editorials filled with vigor and eloquence. It reigned for five weeks.

Politics played an important part in newspapers' content. One side against another, with plenty of fireworks. It could also cost writers their jobs, if as in the following the columnist wrote about some one more powerful.

Joe Goodman of The Territorial Enterprise saw his demise as a result of a "welcome home" article he directed at William Sharon. Sharon, a member of the San Francisco "Bank Crowd" had returned to Virginia CIty in 1872, to pursue his political interests in the United States Senate. Goodman wrote:

". . . Your career in Nevada for the past nine years has been one of merciless rapacity. You fastened yourself upon the vitals of the State like a hyena, and woe to him who disputed with you a single coveted morsel of your prey. . . . You cast honor, honesty, and the commonest civilities aside. You broke faith with men whenever you could subserve your purpose by so doing."

Sharon lost the election that year. In 1874 he entered the same race and bought the newspaper. As with privileges of a new owner, the following article appeared regarding candidate Sharon:

"The present prosperity of western Nevada is more due to him than to any other ten men, and could his work here be stricken out, with it would go at once two-thirds of our people, improvements and wealth."

To the publisher went the spoils!

Reverend Grosch Seeks His Sons' Claim

In the fall of 1863 Reverend A.B. Grosch, father of Allen and Hosea, formed the Grosch Gold & Silver Mining Company. Joining with Rev. Grosch were seven other men. With $10,000,000 they hired the best counsel, and well known citizens of San Francisco to act as trustees. Their claim was to recover the property of Allen and Hosea Grosch who had accidentally died a few years before.

Allen and Hosea were credited with determining that "the blue stuff" was silver. They had further made original claim to the land held by the Gould & Curry, Ophir and Mexican mining companies.

Sandy Bowers and Eilley Orrum knew of the boys' claims, as did Laura Ellis who remembered seeing the book in which the young Grosches had filed all their statistics. The brothers had even pointed out their claim to Ellis, who was able to locate the exact spot.

Joe Goodman, the young editor of The Territorial Enterprise, took another view. To him the claim was absurd. So absurd that Goodman wrote, and printed, nursery jingles about the Grosches. The rhythms delighted his readers, and were no doubt scoffed by the plaintiffs.

Other factors were to play a defining role in the dispute. Had Rev. Grosch known more about the character of Judge George Turner, who would hear the complaint, perhaps the $10,000,000 might have been better spent.

S.L. Davis wrote in *Nevada*, "Judge Turner had developed into the most mercenary man in Washoe."

Money had become so plentiful in Virginia City that men would do most anything for a pound of silver. Judge Turner was such a man.

As it turned out Rev. Grosch soon saw the futility of trying to recover his sons' property. On the advice of counsel, the suit was dropped.

Men Who Unearthed the Millions

Sandy Bowers was one of the first to hit the bonanza jackpot. Every month about $100,000 filled his pockets. Sadly, he and Eilley's claims ran dry.

Old Henry Comstock, a discoverer of everything

Sandy Bowers

in sight, didn't know what he had found - all that "blue stuff"got in the way of his digging. He sold out his share of the Ophir Mine for $11,000. A few years later he put an end to his life with a revolver.

James "Old Virginny" Finney, who discovered the location of the first quartz mining claim on the Comstock, and named the town, sold out for a quart of whiskey and a blind mustang. He died a pauper.

Pat McLaughin and Peter O'Reilly sold their claims for $40,000. McLaughlin went to Montana and became a cook. O'Reilly died in a madhouse and was buried in a pauper's grave.

Alvah Gould, who co- found the Gould & Curry mine, sold his share for $450. He spent it all in one swoop, telling how he had swindled some "suckers."

Men Who Reaped the Millions

The shrewd men who, for meager dollars, bought out the pioneers' claims went on to become the beneficiaries of millions and more millions.They became United States Senators, land barons, and financiers.

John MacKay, toiled up Sun Mountain with a shovel on his back. His discovery and genius gave him the title "Bonanza King" In time he also became a Nevada Senator. Is enduring legacy, was the Mackay School of Mines, University of Nevada.

William Sharon, once a millionaire, and a founder of the Virginia & Truckee Railroad, didn't fair so well. After the Bank of California failed, his body was found in San Francisco Bay.

A LARGE ASSORTMENT OF

MAHOGANY AND METALLIC COFFINS,

LEAD COFFINS, SHROUDS, COLLARS, CRAVATS,

SILVER COFFIN PLATES, GRAVE STONES

ETC., CONSTANTLY ON HAND.

HEARSES and COACHES, together with every variety of FUNERAL EQUIPMENTS, furnished to order.

Especial attention given to DISINTERRING BODIES, and preparing them for Shipment to the Atlantic States and elsewhere.

IRON AND WOOD GRAVE INCLOSURES

—AND—

MARBLE TOMB STONES,

FURNISHED TO ORDER.

The end of one's era

William C. Ralston, was a partner in the V&T Railroad and the Bank of California.

James Fair, invested millions in San Francisco real estate, including the Fairmont Hotel.

Adolph Sutro conceived the extraordinary Sutro Tunnel. He invested his millions in San Francisco real estate, and was once mayor of the city.

Square Set Timbering

Philipp Deidesheimer

Early in the 1860s miners on the Comstock struck ore bodies that were larger and deeper than any that had been found anywhere in the world. This compounded the usual problem found in all mines - how to keep the earth from caving in on the miners. The conventional method - using logs of up to 20 feet in one piece, or longer ones bolted together up to 40 feet to act as posts and caps to hold up the ground, was not workable. Consequently cave-ins resulting in miner's deaths were frequent and terrible.

Philipp Deidesheimer, a German immigrant working at the time in the California mines where he had developed a reputation for innovative engineering, was hired to help solve the timbering problem.

In 1860, after spending only a few weeks in Virginia City analyzing the situation of the deep mines, he came up with the idea of using short timbers of four to six feet one to build "cribs" which would be stacked on top of each other or joined side by side to provide shoring for the mines. The "cribs" could also be filled in with rock debris to make them more stable.

These square set cribs quickly demonstrated their superiority to anything else devised, and by 1861, all the mines of the Comstock had adopted this technique.

Philipp Deidesheimer became the Superintendent of the Burning Moscow Mine; but he never fully realized any great financial gain from his invention even though it became the standard for deep mines around the world.

As Dan De Quille, noted in The Territorial Enterprise, ". . . the mining men of Europe . . . can hardly fine words in which to adequately express their great admiration of it."

T.L. Dawes, Mining on the Comstock, Gold Hill, Nevada, 1877.

Courtesy of the Nevada Bureau of Mines and Geology

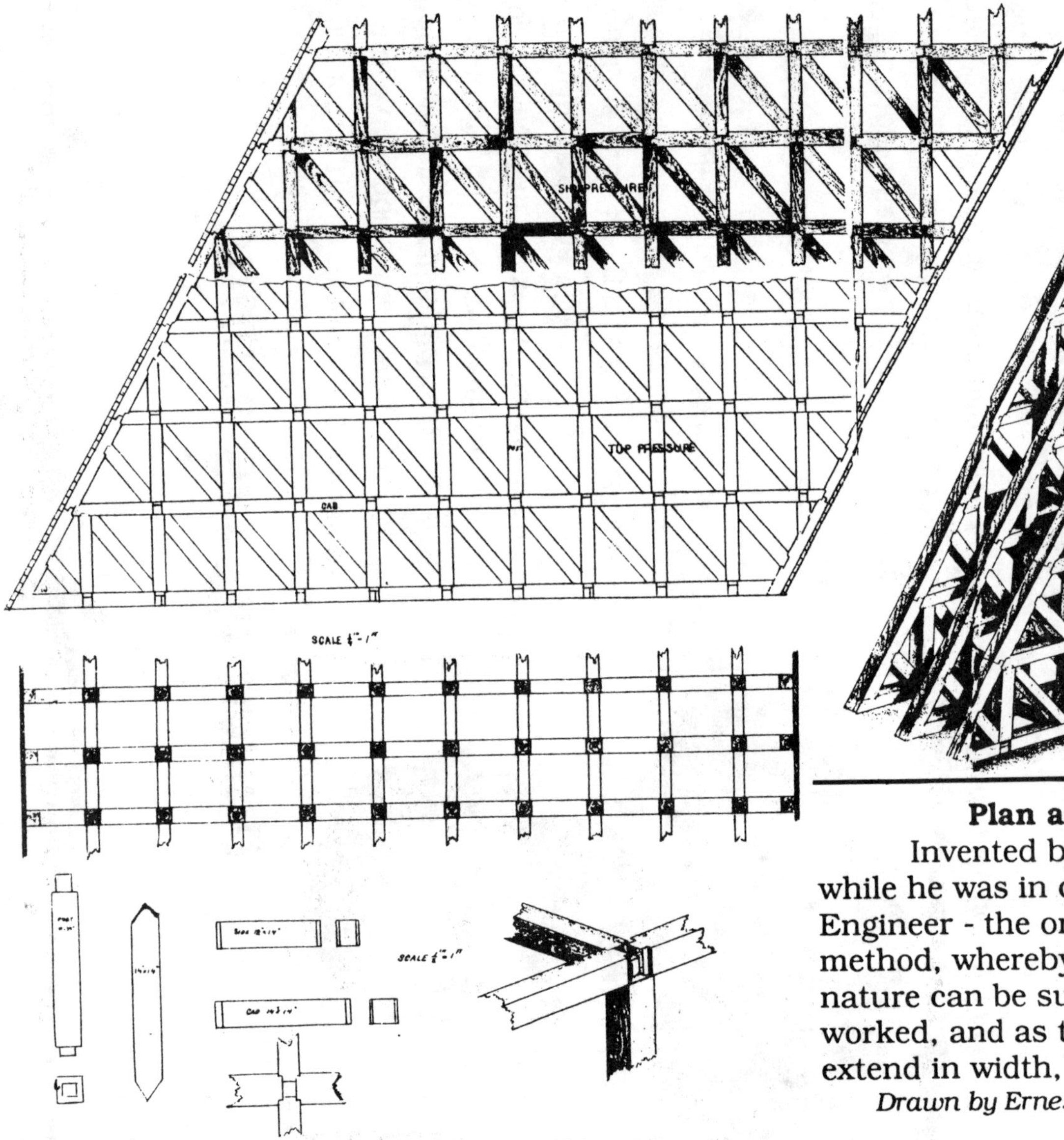

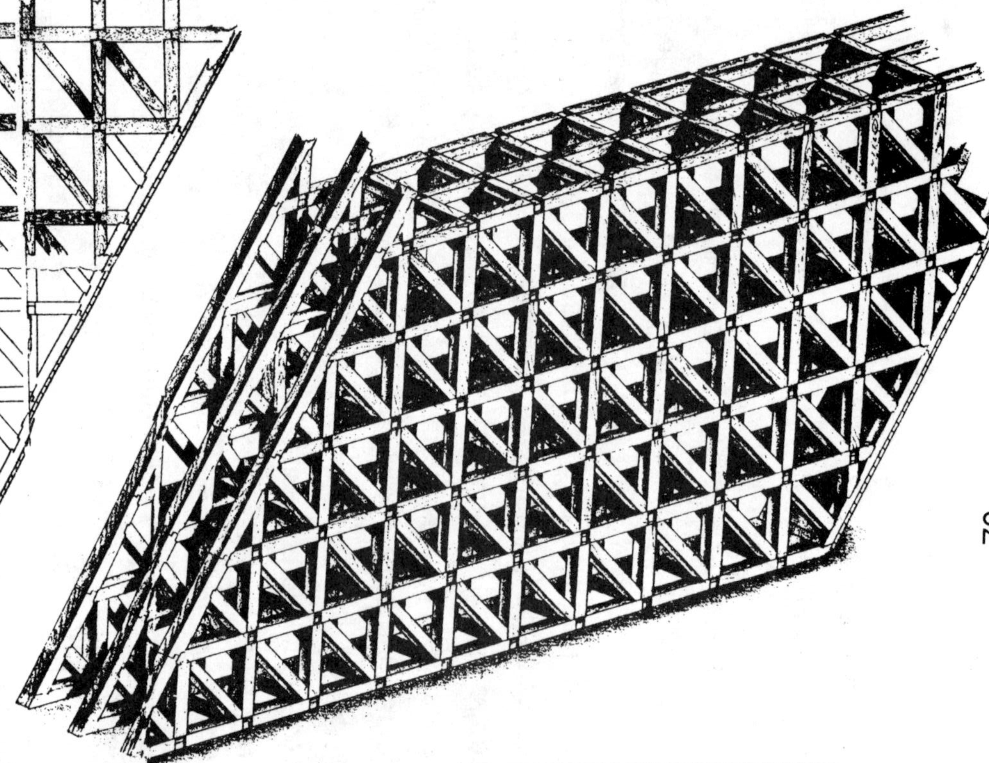

Plan and System of Timbering

Invented by Philipp Deidesheimer in 1860 while he was in charge of the Ophir Mine as Mining Engineer - the only safe, secure and possible method, whereby a lode of mineral of great size or nature can be successfully and economically worked, and as to safety of life and property to any extend in width, depth and length.

Drawn by Ernest Walser, civil engineerr:

San Francisco, April 1875

George Hearst, who rode a mule into Washoe, discovered his treasure then became a United States Senator. Later he gave his son, William Randolph Hearst, the San Francisco Examiner newspaper.

Lucky Baldwin was kicked out of the International Hotel for an unpaid bill, only to become a millionaire soon afterwards. Today, a portion of his estate is the Los Angeles County Arboretum in Arcadia, California.

James Flood and Peter O'Reilly, the names could go on. For it was these hard-driving, far-sighted, and extremely lucky miners that helped build Virginia City, and erect edifices in San Francisco.

Virginia City in 1880

Courtesy Nevada Historical Society

There's gold, and it's haunting and haunting;
It's luring me on as of old;
Yet it isn't the gold that I'm wanting
So much as just finding the gold.

- Robert Service

Acknowledgments

Discovery of the Comstock Lode left its mark in many ways. Thousands of words have been written about the tumultuous events, and the people who made it all happen. Records included in this text are on file in the Wells Fargo History Room, San Francisco; Nevada Historical Society, California Historical Society, California State Library, the Nevada State Library, and Mackay School of Mines, Nevada State University. The author is grateful to the curators of these remarkable institutions, and their meticulously kept archives. Without their diligent and dedicated care, today's readers would know little about their extraordinary past.